A BIOSYSTEMATIC STUDY OF THE GENUS BRODIAEA (AMARYLLIDACEAE)

BY
THEODORE F. NIEHAUS

UNIVERSITY OF CALIFORNIA PUBLICATIONS IN BOTANY
Volume 60

UNIVERSITY OF CALIFORNIA PRESS

A BIOSYSTEMATIC STUDY OF THE GENUS BRODIAEA (AMARYLLIDACEAE)

A BIOSYSTEMATIC STUDY OF THE GENUS BRODIAEA (AMARYLLIDACEAE)

BY

THEODORE F. NIEHAUS

UNIVERSITY OF CALIFORNIA PRESS
BERKELEY • LOS ANGELES • LONDON
1971

University of California Publications in Botany

Volume 60

Approved for publication May 28, 1971
Issued December 29, 1971

University of California Press
Berkeley and Los Angeles
California

University of California Press, Ltd.
London, England

ISBN: 0-520-09390-9

Library of Congress Catalog Card No.: 77-170326

printed in the united states of america

CONTENTS

A BIOSYSTEMATIC STUDY OF THE GENUS BRODIAEA (AMARYLLIDACEAE)

BY

THEODORE F. NIEHAUS

INTRODUCTION

THE natural distribution of the genus *Brodiaea* is restricted to western North America from northern Baja California to Vancouver, British Columbia. *Brodiaea* has been interpreted by some workers (Jepson, 1925; Munz, 1959) as consisting of a single large genus with three subgenera. Other workers believe with me that these three subgenera should be recognized as the three distinct genera *Brodiaea, Triteleia,* and *Dichelostemma.* This problem is discussed in the section on taxonomy. *Brodiaea,* as referred to in this paper, excludes the other two genera or subgenera just mentioned. As thus defined *Brodiaea* consists of fifteen species, all of which occur in California west of the Sierra Nevada. Three species, *B. coronaria, B. elegans,* and *B. terrestris,* extend northward into Oregon and Washington. The plants are perennials which arise from a corm each growing season. The cycle of growth is initiated each fall from a summer-dormant corm. The leaves and later the flowering scape are produced during the fall, winter, and early spring months. In late spring purple-violet to rose-pink flowers are produced on an umbelliferous scape.

The genus *Brodiaea* has presented problems to various workers who have attempted to provide taxonomic treatments for this group. Narrowing the generic concept (as discussed elsewhere) will aid considerably in understanding species relationships, because a compact, homogeneous group of species emerges which can be examined without confusing side-issues emanating from other groups. Through various techniques, such as comparative morphology, chromatography, chromosome cytology, garden hybridization studies, edaphic studies, and pollination ecology, which have been discovered or improved in recent years biosystematists have been able to add much information on problematical groups of plants (Davis and Heywood, 1963). In this study most of the available techniques were employed.

Certain *Brodiaea* taxa recognized by Hoover (1939*b*), a long-time and careful student of the whole group, pose many questions as to their affinity with other species in the genus. Hoover regarded several morphologically dissimilar taxa as varieties of *B. coronaria.* Some of these taxa occur only in small, very local areas of distribution; some grow on serpentine soils and others are found only on non-serpentine soils. One might suspect that such narrowly distributed species are not valid but represent local morphological variation. But if they are valid species, what explains their restriction to very small populations? With this problem in mind, my study was initiated to determine relationships among the species and hopefully to arrive at a better understanding of the taxonomy of *Brodiaea,* and also to obtain some insight into evolutionary processes.

Differences in the morphology of the flowers have provided important distinguishing characters for separating the species of *Brodiaea*. Examination of gross morphology and vascular anatomy of the flower revealed many differences that have been overlooked by previous workers. These additional differences when coupled with other known characters have made possible a better definition of the species and have shown that some proposed taxa do not merit recognition. Chromatographic observations of *Brodiaea* plants revealed that a fairly large number of phenolic compounds are present. Each species has some phenolics which are specific to it and others which are shared with related species.

A few reports of chromosome number were hitherto available for *Brodiaea* but did not always agree with one another (Johansen, 1932; Burbanck, 1941). This problem demanded a thorough survey of chromosome numbers and karyotypes. My results show that some of the previous chromosome counts were in error and I found that polyploidy is present in most species. An interspecific hybridization program was aimed at determining fertility relationships among species. The large flowers and the ease of emasculation made interspecific pollination appear to be relatively easy. The drawback to such a program is that two or more years are required to grow a flowering plant of *Brodiaea* from seed.

Insect pollinators were observed and collected on all species of *Brodiaea* in the hope of determining the principal means of pollination and whether interspecific pollination occurs. Observations of frequent insect visitors to *Brodiaea* showed that Lepidoptera and Diptera are not effective pollinators; Coleoptera and Hymenoptera are the important pollinators. It is possible for interspecific pollination to occur, but such pollinations are not followed by establishment of hybrids, and many field observations indicate that interspecific hybrids are rare.

Early in the study it was noticed that many of the taxa appear to grow on distinctive soils. An attempt is made to relate taxa to soil preferences, although the information on soil types is incomplete since only areas of economic value have been mapped.

Information from all the above-mentioned areas of research was combined to arrive at a possible explanation of the relationships within each of the various clusters of species in *Brodiaea* as well as their relationship to other clusters in the genus. A natural byproduct of this study is a new taxonomic treatment for the genus.

The genus *Brodiaea* as circumscribed in this paper consists of:

1. *B. appendiculata* Hoover
2. *B. californica* Lindl.
3. *B. leptandra* (Greene) Baker

4a. *B. coronaria* (Salisb.) Engler subsp. *coronaria*
4b. *B. coronaria* subsp. *rosea* (Greene) Niehaus
5a. *B. elegans* Hoover subsp. *elegans*
5b. *B. elegans* subsp. *hooveri* Niehaus
6a. *B. terrestris* Kell. subsp. *terrestris*
6b. *B. terrestris* subsp. *kernensis* (Hoover) Niehaus

7. *B. jolonensis* Eastw.
8. *B. stellaris* S. Wats.
9. *B. insignis* (Jepson) Niehaus

10. *B. pallida* Hoover
11. *B. minor* (Benth.) S. Wats.
12. *B. purdyi* Eastw.
13. *B. filifolia* S. Wats.
14. *B. kinkiensis* Niehaus
15. *B. orcuttii* (Greene) Baker

Acknowledgments

I wish to thank Robert Ornduff, Lincoln Constance, and G. L. Stebbins, Jr., for guidance and useful criticisms during the course of this study. I would also like to thank Glenn Keator, who recently completed a study of *Dichelostemma,* for freely giving advice and communicating observations he has made on *Brodiaea* and related genera; Rimo Bacigalupi for assistance and advice; Wayne Roderick for many specimens and invaluable help in growing *Brodiaea;* Samuel Pusateri for information and hospitality while I was collecting in Tulare County; Ann Rathbun for aid and information while I was collecting in Butte County; Ernest Twisselman for contributing information and collections from Kern County; the United States Navy for transportation and assistance in visiting the Channel Islands; and Charlotte Mentges for supplying many of the illustrations and offering advice on the preparation of others.

The curators of the following institutions have generously allowed me to study their material either in those institutions or on loan, an opportunity without which this study would have been impossible.

BM	British Museum, London
CAS	California Academy of Sciences, San Francisco
CGE	University of Cambridge, Cambridge, England
DAV	University of California, Davis
DS	Dudley Herbarium, Stanford University
GH	Gray Herbarium, Harvard University, Cambridge
JEPS	Jepson Herbarium, University of California, Berkeley
K	Royal Botanic Gardens, Kew
LA	University of California, Los Angeles
LAM	Los Angeles County Museum, Los Angeles
ND	University of Notre Dame, Notre Dame, Indiana
ND-G	Greene Collection, University of Notre Dame
NY	New York Botanical Garden, New York City
OBI	California State Polytechnic College, San Luis Obispo
ORE	University of Oregon, Eugene
OSC	Oregon State University, Corvallis
RSA	Rancho Santa Ana Botanic Garden, Claremont, California
SBBG	Santa Barbara Botanic Garden, Santa Barbara, California
SBM	Santa Barbara Museum
SD	San Diego Museum of Natural History, San Diego, California
UC	University of California, Berkeley
US	Department of Botany, United States National Herbarium, Smithsonian Institution, Washington
USC	University of Southern California, Los Angeles
WILLU	Willamette University, Salem, Oregon
WS	Washington State University, Pullman
WTU	University of Washington, Seattle

MORPHOLOGY

Corm

The corm is the principal means by which plants of *Brodiaea* perpetuate themselves from one growing season to another. Dormancy begins in early summer when the corm has become an enlarged organ filled with starch. The stored starch enables the plant to commence rapid growth in the fall when dormancy is broken.

At the beginning of the growing season in September the corm shows activity in its apical region prior to the first fall rains. The leaves begin to elongate, often to 4 or 5 mm, and root tips protrude 1–2 mm from the basal root plate. Further growth occurs after the first heavy rainfall in which rainwater penetrates to the level of the corm. This has been observed in nature as well as under controlled conditions. A new growing cycle is thus activated, with the roots growing 1–2 cm in the first 24 hours. The roots are 5–8 cm long after the first four or five days. The leaves continue to elongate but at a slower rate than the roots. The starch stored in the corm is rapidly metabolized by these growth activities and decreases in volume; the corm becomes severely withered after several months. A new corm begins to form at the base of the new leaves about a month after the growing season starts. It is mature before flowering and usually occupies the same position as the previous corm. At maturity of the new corm the only remnant of the old one is a thin membranous layer of dead tissue under the new corm.

The new corm of mature plants often produces two to fifteen smaller cormlets adjacent to it. Hoover (1939*b*), reporting that some members of the genus did not reproduce by means of offsets, used this as an important reason to differentiate some species. However, my observations indicate that these cormlets are produced by all members of the genus; hence their presence is of no diagnostic value in differentiating any of the species. The cormlets are produced in the axils of the old leaf bases on the mature corm. This has also been observed by Smith (1930) for *Triteleia hyacinthina*. The cormlet generally will not produce a flowering plant until one or more growing seasons have ensued, during which time the cormlet enlarges to mature size adjacent to the parent plant. In addition to the adjacent cormlets the parent plant may produce cormlets which are dispersed by contractile roots (plate 1). A fleshy contractile root develops which draws the cormlet 10–15 cm laterally from the parent corm. A contractile root always occurs in seedlings, but such roots serve to draw the first-year cormlet 5–10 cm below the soil surface during the first year and to a greater depth the following season. Once the corm enlarges to adult size its movement by a contractile root ceases.

Extensive studies on the morphological nature of corms have been carried out by Geiger (1929) on Gladiolus and by Smith (1930) on *Triteleia hyacinthina*. They concluded that the corm in these two genera is a highly contracted subterranean stem as ordinarily defined. Nodes are still evident and the modified leaf bases are attached to these nodes. The corm of *Brodiaea* varies from 0.5 cm to 4.0 cm in diameter. Corm size generally correlated with overall plant size, but is related also to site and seasonal conditions. The corms of the smaller species are usually found at a shallower depth of soil than those of the larger species. Collections of corms over the ranges of the various species and during different growing seasons have shown so much variation in size that further studies were abandoned as unproductive. The corms of *B. elegans* alone span the entire size range of 0.5 cm to 4.0 cm in diameter.

Leaf

The leaves of *Brodiaea* begin growth in late summer before the first heavy rainstorm. Leaf growth is slow at first, probably because of low winter temperatures. During February and March, when temperatures become higher, the leaves reach their full length. A slight elongation may occur until late April or May, when growth ceases and senescence ensues; at this time flowering and seed-set occur. The leaves of all species were observed to be dead or nearly so at the time of flowering.

The number of leaves per plant was counted to determine if it was constant for each species. The number of leaves per plant in all species varies from two to six, with three being the average. The sole exception is *B. kinkiensis,* which has only one. Three patterns of leaf habit were observed in the genus: leaves of *B. kinkiensis, B. orcuttii,* and *B. pallida* are erect; those of *B. coronaria* subsp. *rosea* are decumbent; and leaves of all other species are ascending, that is, intermediate in position between erect and decumbent.

The leaf consists of two distinct parts: an expanded basal portion which surrounds the corm and is referred to as the corm coat (Hoover, 1939*a,* 1939*b;* Moore, 1953; Munz, 1959), and the aerial portion, referred to as the leaf proper (discussed below).

The morphology of the corm coat was used by Hoover (1939*a*) to differentiate the genera *Brodiaea, Dichelostemma,* and *Triteleia.* Moore (1953) pointed out that within subtribe Brodiaeinae the genera occurring in Mexico (e.g., *Dandya, Bessera,* and *Milla*) can be separated from the California genera (e.g., *Brodiaea, Dichelostemma,* and *Triteleia*) on the basis of the parallel fibers in corm coats in the former and the reticulate pattern in the latter. Keator (1967), who also examined the corm coat (in subtribe Brodiaeinae) for differences between California genera, decided that there were no significant differences in fiber pattern.

In *Brodiaea* the dried corm coat consists of a definite reticulate system of fibers with a thin papery sheet of epidermal cells in the areoles (i.e., the region between fibers). The fibers are very heavy and coarse in *B. californica, B. leptandra,* and *B. kinkiensis* in comparison with the other members of the genus. According to Keator (1967), the pattern and shape of the parenchymatous cells between the vascular strands of the corm coat are of taxonomic value in delimiting the genera *Brodiaea, Dichelostemma,* and *Triteleia.* He reports that the shape of parenchyma cells differs between genera and is constant within each genus. His figures seem to be typical of parenchyma cells found in most genera of monocotyledonous plants. I suspect that Keator was comparing cell shapes from different depths of leaf-base tissue, for I have found that the epidermal and parenchymatous cells between vascular strands of the corm coat are similar for all species of *Brodiaea, Dichelostemma,* and *Triteleia* and are thus of no taxonomic value in distinguishing these genera.

The structure of the aerial portion of the leaf is essentially the same in all members of *Brodiaea.* The external wall of the epidermal cells is cutinized and characteristically has projecting ridges (fig. 1) that are more prominent on the adaxial surface. The number of ridges is generally three per cell on the abaxial surface and varies from none to three on the adaxial surface (table 1). The number of ridges on an epidermal cell can be used to separate closely related species: *B. filifolia* has one ridge, *B. kinkiensis* has two, and *B. orcuttii* has three.

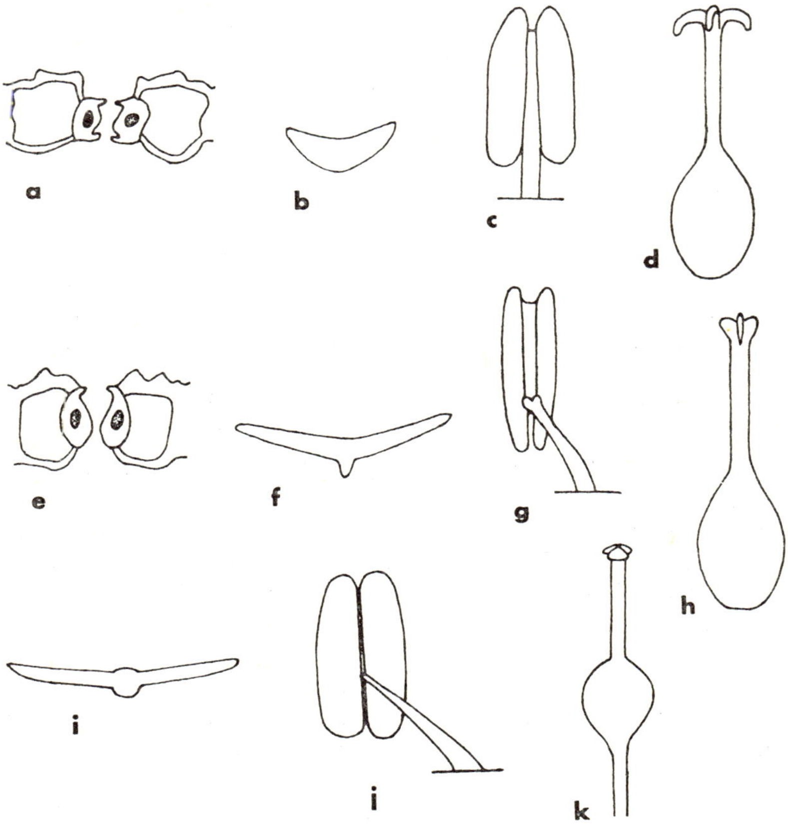

Fig. 1. Morphology of the genera *Brodiaea: a,* x.s. epidermis at guard cell; *b,* x.s. leaf; *c,* stamen; *d,* ovary and stigma. *Dichelostemma; e,* x.s. epidermis at guard cell; *f,* x.s. leaf; *g,* stamen; *h,* ovary and stigma. *Triteleia: e,* x.s. epidermis at guard cell; *i,* x.s. leaf; *j,* stamen; *k,* ovary and stigma.

In *Brodiaea* the guard cell is sunken and has two sets of ledges, that is, a hornlike projection (Esau, 1953), whereas in *Dichelostemma* and *Triteleia* they are not sunken and have only an upper set of ledges (fig. 1).

Internal to the epidermis are one to three layers of palisade parenchyma; the number of these layers is generally the same on both abaxial and adaxial surfaces. Internal to the palisade parenchyma, the center of the leaf is composed of spongy parenchyma. Vascular bundles are scattered around the leaf in the zone between palisade parenchyma and spongy parenchyma. The number of vascular bundles varies from species to species, depending in part on leaf size (table 1).

The outline of the leaf in transection has been used by Hoover (1939*a*), Yausy (1942), and Keator (1967) to differentiate *Brodiaea, Dichelostemma,* and *Triteleia*. *Brodiaea* has a rounded abaxial surface, whereas the other two genera have a central abaxial keel. The transection of the leaf of many taxa of *Brodiaea* is crescent-shaped or, as in *B. kinkiensis,* terete. The epidermal cells of leaves in *Brodiaea* vary considerably in length and width.

TABLE 1

Number of Epidermal Ridges, Palisade Layers, and Vascular Bundles Found in Brodiaea

Species	Epidermal ridges per cell		Palisade layers		Vascular bundles per leaf[a]
	Abaxial	Adaxial	Abaxial	Adaxial	
B. appendiculata	0	3	2	2	33
B. californica	0	3	2	2	31
B. leptandra	0	1	2	2	18
B. coronaria	0	3	2	2	23
B. coronaria subsp. *rosea*	0	1	2	3	18
B. terrestris	2	3	2	2	18
B. terrestris subsp. *kernensis*	3	3	3	2	18
B. elegans	2	3	2	1	20
B. elegans subsp. *hooveri*	2	3	2	2	38
B. filifolia	0	1	2	2	18
B. kinkiensis	2	3	3	2	29
B. orcuttii	3	3	2	3	20
B. jolonensis	0	3	1	2	22
B. stellaris	3	3	1	1	13
B. pallida	0	3	1	1	30
B. minor	0	3	2	2	14
B. purdyi	0	3	2	2	16
B. insignis	0	3	1	2	24

[a] A leaf from five plants of each species locality listed in the appendix was examined.

Inflorescence and Flower

The pedicels borne at the apex of the scape are subtended by scarious bracts, as is typical for all members of Amaryllidaceae. The scape grows from the apex of the corm and first appears at the soil surface in January, three to four months after the start of the growing cycle. Scape, pedicels, and floral buds remain small until March, when the scape begins to elongate at its base, where the cells elongate and become rigid. The upper part of the scape matures last. This is followed by elongation of pedicels and by flowering. Superficially the scape in *Brodiaea* appears to be terete, but at the microscopic level it is polygonal in transection, as are the pedicels. The central part of the scape is hollow. The mature scape varies considerably in length from one species to another (see table 2). Length of scape was found to vary according to the amount of moisture available during the growing season. In drier years scape length may be reduced by 50 percent or more of normal. The apex of the scape is enlarged where the pedicels and their subtending bracts are attached. The outermost bract surrounds the scape apex, and a smaller bract is borne at the base of each pedicel. The number of nerves in the outermost bract varies from three or four in *B. minor* to eight to ten in *B. appendiculata*. The number of nerves in the inner bracts varies from one to a few in all species and thus is not of diagnostic value. The longest pedicels are found in *B. terrestris* and *B. appendiculata;* the shortest, in *B. minor* and *B. purdyi*. The pedicels are joined to the scape by an abscission zone. The zone is

TABLE 2

CHARACTERS OF THE SCAPE IN BRODIAEA[a]

Species	Scape length (dm)	Pedicel length (cm)	Nerves in outer bract
B. appendiculata	1–4	4–10	8–10
B. californica	2–7	2–10	8–10
B. leptandra	2–4	2–5	3–6
B. coronaria	0.5–5	1–5	4–8
B. coronaria subsp. *rosea*	0.5–0.7	1–4	5–9
B. elegans	1–5	0.5–10	5–9
B. elegans subsp. *hooveri*	1–5	0.5–10	6–9
B. terrestris	0.5–7	3–15	5–9
B. terrestris subsp. *kernensis*	0.2–2	1–7	5–8
B. jolonensis	0.5–1.5	1–4	9–10
B. stellaris	0.2–0.6	1–5	7–8
B. pallida	1–2	0.5–3	3–6
B. insignis	0.7–2.5	2–9	5–7
B. minor	0.2–1	1–3	3–4
B. purdyi	1–2.5	1–3	3–4
B. filifolia	2–3	1–4	4–8
B. kinkiensis	2–3	3–8	4–9
B. orcuttii	0.8–2.5	1–5	4–6

[a] Range of variation measurements are from all the herbarium specimens examined in this study, including some which are duplicates representing both "dry" and "normal" growing seasons.

relatively tenacious in most species, but in *B. orcuttii* and *B. filifolia* it is so weak that slight pressure will detach the pedicels during flowering. In other species the abscission layer does not mature until the seeds are ripe in late summer or early fall.

The perianth consists of three narrow outer segments and three broader inner segments. The lower portions of all six are fused into a light-green tube, with stripes of violet-purple continuing toward the base from the free portion of the perianth. The perianth is variable between species, as may be seen in the illustrations for each species

Staminodia occur opposite the three outer perianth segments. These structures are flat, horn-shaped, slightly hooded, and short-lanceolate, or may be lacking in some species. In position the staminodia are either appressed, leaning inward, or distant from the stamens. These staminodial differences can be seen in the illustrations for each species.

The three stamens opposite the inner perianth segments provide many taxonomically useful morphological characters. The filaments of all species are linear. The abaxial side of the free portion of the filament has a thin triangular flap in some species, but this flap is lacking in others. The abaxial surface of the apical part of the filament may be rounded or slightly bilobed. In *B. stellaris* it is elongated into two large appendages angel-winged in appearance. Two narrow elongate filiform appendages are occasionally found in *B. californica.* The free portion of the filament may be either short or long in different taxa. Anthers vary from lanceolate to oblong. The basal lobes of the two pollen sacs are essentially similar in shape in all taxa, but the apical lobes differ in form. The notch between the lobes may be narrowly rounded, broadly rounded, or broadly obtuse. The tips of the pollen sac are erect or sharply hooked inward. The base

TABLE 3

Condition of Perianth Tube on Mature Capsules of Brodiaea

Species	Splitting characteristic	Membrane thickness	Membrane transparency
B. appendiculata	Longitudinal	Thick	Clear
B. californica	Longitudinal	Thick	Clear
B. leptandra	Longitudinal	Thin	Clear
B. coronaria	No	Thick	Opaque
B. coronaria subsp. *rosea*	No	Thick	Opaque
B. elegans	No	Thick	Opaque
B. elegans subsp. *hooveri*	No	Thick	Opaque
B. terrestris	No	Thick	Opaque
B. terrestris subsp. *kernensis*	No	Thick	Opaque
B. jolonensis	No	Thick	Opaque
B. stellaris	No	Thin	Clear
B. pallida	No	Thick	Opaque
B. insignis	No	Thick	Opaque
B. minor	No	Thick	Opaque
B. purdyi	No	Thick	Opaque
B. filifolia	Longitudinal	Thin	Clear
B. kinkiensis	No	Thick	Opaque
B. orcuttii	Longitudinal	Thin	Clear

of the notch in some taxa has a dentate appendage produced by elongation of the anther connective. The connective tissue of the anther may or may not be deeply channeled on the abaxial surface. The abaxial tissue usually has a smooth appearance, but a number of taxa have enlarged cells, which in *B. pallida* are so strikingly enlarged that Hoover (1938) called them "processes" and used their prominence as a character to differentiate this species from all others.

The gynoecia of *Brodiaea* species are quite similar morphologically. The ovary may be ovoid, obovate, or globose. The ovary consists of three loosely adnate carpels with the ovules borne on an axial column. The ovary is green in all species except *B. jolonensis,* in which it is purplish. Styles vary in length from 0.4 cm to 2.3 cm. The ratio of style length to ovary length is of taxonomic value in differentiating many species. The stigma consists of three flattened wings with papillae along their upper surface. The fruit is a loculicidal capsule which matures in mid- or late summer. The scape and pedicels enter senescence a month or more before the capsule matures. The perianth tube surrounding the capsule dries into a thin or thick membrane which may be either hyaline or opaque, and in some species splits as the capsule enlarges (table 3). The nature of the membrane and whether it splits are constant within each species.

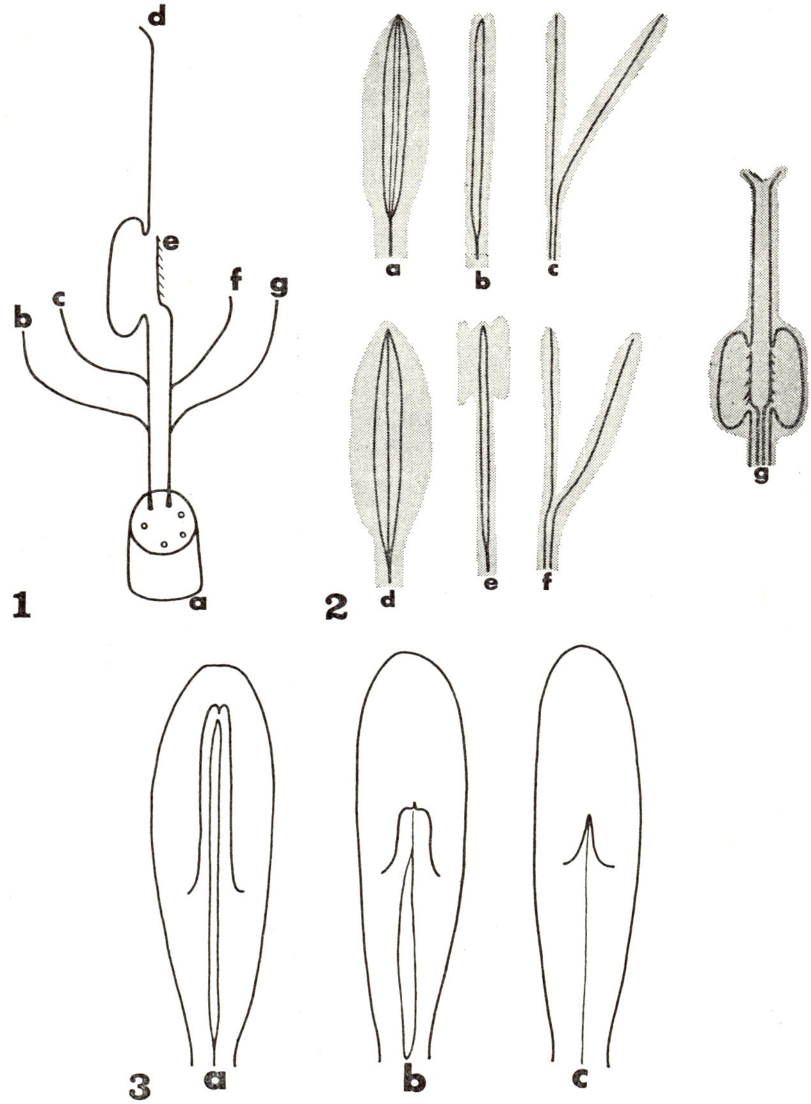

Fig. 2. 1. Vascular anatomy of a *Brodiaea* flower. Two vascular bundles of six that enter flower are shown: *a,* pedicel; *b.* outer perianth segment; *c,* staminodium; *d,* style; *e,* ovules; *f,* stamen; *g,* inner perianth segment. 2. Vasculature most commonly found in *Brodiaea* (variations are noted in table 4): *a,* outer perianth segment; *b,* staminodium; *c,* vertical sagittal section of *a* and *b* as found in the perianth; *d,* inner perianth segment; *e,* stamen; *f,* vertical sagittal section of *d* and *e* as found in the perianth; *g,* gynoecium. 3. Vascular bundles of staminodia in *Brodiaea: a,* as found in majority of species; *b, B. kinkiensis; c, B. filifolia.*

Characterization of seed shape for *Brodiaea* has been attempted by Hoover (1939*a*) and Keator (1967) with slight success. Angularity of the seed makes it difficult to define the seed in a few precise terms. Seeds are approximately elliptical, with three or four sharply angled sides. The sclerid pattern of the seed coat is, however, very characteristic for *Brodiaea.* It has been described as longitudinally striate by both Hoover (1939*a*) and Keator (1967). I have found that seed coats of all species of *Brodiaea* are made up of longitudinal sclerids. Size and shape of the plates are similar in all species. The slight variations observed are probably due to growth conditions. On the basis of observation using the scanning electron microscope the seed-coat surface in *Brodiaea* is of elongated sclerids; in *Dichelostemma,* raised rectangular to hexagonal; in *Triteleia,* raised rectangular (plate 1).

The vascular anatomy of the flower was studied by clearing with NaOH and staining with saframin (Foster, 1949). The pedicel immediately below the flower has six large vascular bundles. Three of these are somewhat larger than the others and supply the outer three perianth segments. Each of the three larger vascular bundles has two lateral branches which supply the outer perianth segment and the staminodium, respectively (fig. 2). The main axis of the outer vascular bundle continues into the ovary around the outer dorsal wall, ending at the tip of the style. The three smaller vascular bundles supply the inner perianth segment and the stamen, respectively. The main axis of the smaller vascular bundle continues into the central column of the ovary and supplies the ovules in one of the three locules. The vascular bundle of the outer perianth segment divides into three strands near the base of the perianth segment. The two outer lateral bundles run to the tip of the perianth segment as single, double, or triple strands. The central bundle divides into two or three strands which run to the tip of the segment, where all the strands fuse together. The staminodial vascular bundle divides into two branch strands near its origin. These extend to the tip of the staminodium, where the branch strands then fuse. The majority of species have such double staminodial strands.

Interestingly, the staminodial vascular bundles in three species exhibit various degrees of fusion, reduction, or loss of parts (e.g., *B. kinkiensis, B. filifolia, B. orcuttii*), (fig. 2). *Brodiaea kinkiensis* shows a partial reduction of the staminodia themselves. The vascular bundles are double for much of their length along the perianth member, but become reduced in size in the upper part of the staminodium. The staminodium in *B. filifolia* is reduced to a short filiform flap of tissue; the staminodial strand of this species is reduced to a single bundle which is one cell or a few cells in thickness and is difficult to detect in stained flowers. In *B. orcuttii* there are three stamens, as in all other *Brodiaea* species, and basal tissue along the perianth segment, but there are no staminodia. The staminodial vascular bundles are absent and were not detected even as short branches from the main vascular strand in *B. orcuttii.* The vascular bundle of the inner perianth segment divides into three strands near its base. These strands continue to the tip, where they fuse; each of the three main strands may or may not divide further into two or three additional branch strands. The vascular bundle of the stamen divides into two strands which reunite at the apex.

Variation in number of vascular bundles in both inner and outer perianth segments can be used to differentiate species (table 4). The evolutionary significance of the different numbers of vascular bundles in the staminodia and perianth segments are discussed in the section on evolution.

TABLE 4

VASCULAR STRANDS IN PERIANTH SEGMENTS OF BRODIAEA

Species	No. of strands in outer segment	No. of strands in inner segment
B. appendiculata	1–2–1	1–1–1
B. californica	1–2–1	1–1–1
B. leptandra	1–2–1	2–2–2
B. coronaria	1–2–1	1–2–1
B. coronaria subsp. *rosea*	1–2–1	1–2–1
B. elegans	3–3–3	3–3–3
B. elegans subsp. *hooveri*	3–3–3	3–3–3
B. terrestris	1–2–1	1–1–1
B. terrestris subsp. *kernensis*	1–2–1	3–3–3
B. jolonensis	1–2–1	1–1–1
B. stellaris	2–0–2	2–0–2
B. pallida	1–2–1	2–1–2
B. insignis	1–2–1	1–2–1
B. minor	1–2–1	1–2–1
B. purdyi	1–2–1	1–2/1–1
B. filifolia	1–2–1	1–1–1
B. kinkiensis	1–2–1	2–2–2
B. orcuttii	1–2–1	1–2–1

POLLEN

A number of preparatory techniques were tried before the pollen of *Brodiaea* was studied. The aniline oil and stain technique ordinarily used for monocotyledonous pollen preparations (Cranwell, 1953) was unsatisfactory. The pollen tended not to stain; even when an adequate stain was obtained, it was difficult to see surface characteristics. This problem was overcome by using the acetolysis method and by mounting the pollen with glycerin jelly (Erdtman, 1952). All measurements reported here are from acetolyzed pollen. A sample of fifty pollen grains was measured for each population examined. The measurements for each sample of fifty pollen grains fell within the expectations of a normal curve at the 95 percent confidence level (table 5). Pollen was also examined under the scanning electron microscope. The photographs taken on this microscope clearly show the external morphology (plate 1). Pollen in *Brodiaea* can be characterized as monosulcate with a reniform shape. The grain is smooth, and many small pores are distributed over the entire surface. The same general morphology has been observed in many other genera of tribe Allieae (Erdtman, 1952) in Amaryllidaceae. There are slight differences in angularity of polar ends among the pollen grains of this tribe. Polar ends in *Brodiaea* and *Dichelostemma* are rounded, but they are sharply pointed in *Triteleia* and *Bloomeria*. Pollen size varied according to the level of polyploidy of the species or race. An increase in chromosome number of a given species resulted in correspondingly larger pollen grains. However, pollen size cannot be used alone as a criterion for distribution of species, since diploid pollen of one species (e.g., may be larger than that of a tetraploid species *B. appendiculata* 50.4μ *n* = 6, versus *B. filifolia* 48.5μ n = 12).

TABLE 5

MEAN MEASUREMENTS OF POLLEN OF BRODIAEA[a]

Species	Length	Width	Ratio	Index	Chrom. no.	Coll. no.
B. appendiculata	50.46	35.89	1.40	42.55	n-6	515
B. appendiculata	46.84	35.70	1.38	40.89	n-6	635
B. californica	66.16	44.98	1.47	54.55	n-6	654
B. californica	71.83	50.41	2.90	53.03	n-6	244
B. leptandra	68.73	46.12	1.49	56.03	n-12	679
B. leptandra	60.69	48.12	1.26	54.03	n-6	UCBG[b]
B. coronaria	51.17	40.41	1.27	45.46	n-12	634
B. coronaria	45.36	32.13	1.41	38.17	n-6	706
B. coronaria subsp. *rosea*	42.08	29.46	1.43	35.20	n-6	556
B. elegans	71.21	44.74	1.57	56.43	n-16	695
B. elegans	64.16	39.36	1.59	50.24	n-16	522
B. elegans	61.21	39.51	1.63	49.17	n-8	552
B. elegans subsp. *hooveri*	66.40	41.60	1.60	52.55	n-20	696
B. terrestris	49.27	35.08	1.40	41.58	n-18	674
B. terrestris	45.79	33.13	1.38	38.94	n-6	364
B. terrestris subsp. *kernensis*	59.83	39.46	1.52	48.58	n-24	743
	60.50	36.41	1.54	48.81	n-24	632
B. jolonensis	48.36	34.46	1.40	40.82	n-18	509
B. jolonensis	41.65	27.51	1.51	33.84	n-6	511
B. stellaris	43.03	30.08	1.43	35.96	n-6	673
B. pallida	46.03	33.03	1.39	38.98	n-6	526
B. insignis	53.07	34.46	1.54	42.77	n-18	754
B. minor	43.93	30.23	1.45	36.44	n-6	517
B. minor	39.70	27.04	1.47	32.75	n-6	655
B. purdyi	61.36	32.61	1.52	44.72	n-12	528
B. purdyi	55.64	36.70	1.45	45.17	n-24	648
B. filifolia	48.50	30.99	1.57	38.77	n-12	661
B. kinkiensis	57.12	35.60	1.60	45.10	n-16	407
B. orcuttii	50.41	34.89	1.44	41.94	n-12	663

[a] All measurements are in microns. Ratio is between length and width. Index is square root of length times width (Ting, 1966).
[b] Univ. Calif. Botanical Garden.

COMPARATIVE CHEMISTRY

Chromotographic techniques have been used to determine the number and kinds of compounds in plant materials (Geissman, 1962; Alston and Turner, 1963; Harborne, 1964; Swain, 1966). This information has served to elucidate taxonomic relationships in a number of plant groups (Ellison et al., 1962; Harvey and Grant, 1965; Scora, 1967).

I have studied the flavonoids of *Brodiaea* in the hope of finding clues to the relationships between species or clusters of species. Preliminary chromatograms

TABLE 6

FLAVONOID COMPOUNDS FOUND IN BRODIAEA

Species	Flavonoid compounds																										
	1	2	3	4	5	6	7	8	9	10	11	12	13	14	15	16	17	18	19	20	21	22	23	24	25	26	27
B. leptandra	×	×	×	×	×	×	×	×	×		×	×	×		×			×				×					
B. californica	×	×	×	×	×	×	×	×	×		×	×	×					×				×					
B. appendiculata	×	×	×	×	×	×	×		×		×	×	×	×				×				×					
B. coronaria	×	×	×	×	×	×	×	×	×	×	×	×		×	×	×	×			×							
subsp. *rosea*	×	×	×	×		×	×	×	×	×				×	×		×						×				
B. elegans	×	×	×	×	×	×	×	×	×	×					×	×	×		×								
subsp. *hooveri*	×	×	×	×			×	×	×	×	×	×		×	×				×	×		×		×			
B. jolonensis						×					×		×	×		×					×			×			
B. terrestris	×	×	×	×	×	×	×	×	×	×		×		×	×					×			×				
subsp. *kernensis*	×	×	×	×		×	×	×	×	×	×						×			×							
B. stellaris	×	×	×	×	×	×			×		×		×	×							×	×		×			
B. pallida	×	×	×	×	×	×	×		×		×		×	×	×				×								
B. minor	×	×	×	×	×	×		×		×	×		×		×				×								
B. purdyi	×	×	×	×	×	×		×		×	×	×	×		×				×								
B. insignis	×	×				×	×	×		×						×		×					×		×		
B. kinkiensis	×	×	×	×	×	×	×	×	×	×	×	×	×													×	×
B. filifolia	×	×	×	×	×		×	×	×	×		×		×													
B. orcuttii	×	×	×	×	×		×	×	×	×				×		×	×			×							

were run using the plant parts of *Brodiaea* such as leaves, scape, or flowers to determine if one or more of these parts contained a higher number or a different complement of flavonoids. Initial chromatograms of leaves or scapes were poor in flavonoid compounds; those of flowers were rich in such compounds. Therefore in all subsequent studies extracts of whole flowers were used. The number of flowers required for the best representation of flavonoid compounds per chromatogram was determined by increasing the number of flowers per chromatogram until compounds began merging into an undecipherable jumble. Four flowers per chromatogram proved to be best for all species except one: in *B. elegans* one flower is adequate and two flowers result in a blurred chromatogram. Each population was sampled for two or even three succeeding years to determine if any annual variation in the amount or number of chemical compounds occurred. All chromatograms from the same population were found to be identical, suggesting that seasonal factors do not influence chemical composition. Samples taken from different areas within the distribution of each species always gave similar chromatograms, indicating that there is no intraspecific geographic variation in flavonoid compounds. Chromatograms of plants of *B. elegans* with different ploidy levels were also found to be similar.

Each sample was extracted in methanol for 24 hours. The extract was then spotted on 3MM Whatman chromatographic paper. The first dimension of the chromatogram was run using a solvent consisting of tertiary butyl alcohol, acetic acid, and water in a 3:1:1 ratio. The second dimension was run with 15 percent acetic acid. The chromatogram was then placed over an ultraviolet lamp and the position and color of the fluorescent spots were recorded. Ammonia vapor was applied to each spot and the color change (if any) was noted. A total of twenty-seven spots representing as many different flavonoid compounds was found in the genus *Brodiaea* (table 6). The number of compounds in a given species varied from seven (*B. jolonensis*) to seventeen (*B. coronaria*). Closely related species, such as *B. appendiculata, B. californica,* and *B. leptandra,* have a similar array of flavonoid compounds differing by only one or two compounds; the morphologically more distinct species have a difference of five to ten compounds. However, the three subspecies which are recognized in this paper and which are nearly similar morphologically have three to five spot differences.

The biochemical relationships of each species of *Brodiaea* to all other species of the genus is shown in a polygonal diagram of paired-affinity indices as used by Ellison et al. (1962) and Harvey and Grant (1965). The greater the conformity in area and shape of the polygonal representation between species, the closer their biochemical relationship (Scora, 1967). The affinity indices are determined by the following formula:

$$\frac{\text{Spots in common of taxon } 1+2}{\text{Total spots of taxon } 1\times 2}\times 100.$$

The resulting indices are expressed along the radii from 0 to 100 percent, beginning with 0 percent at the center. The radius representing the species under consideration is always 100, since it is not compared with other species. The species of *Brodiaea* are arranged in the polygonal-affinity diagram with the taxa most similar morphologically next to each other (fig. 3). When the affinity in-

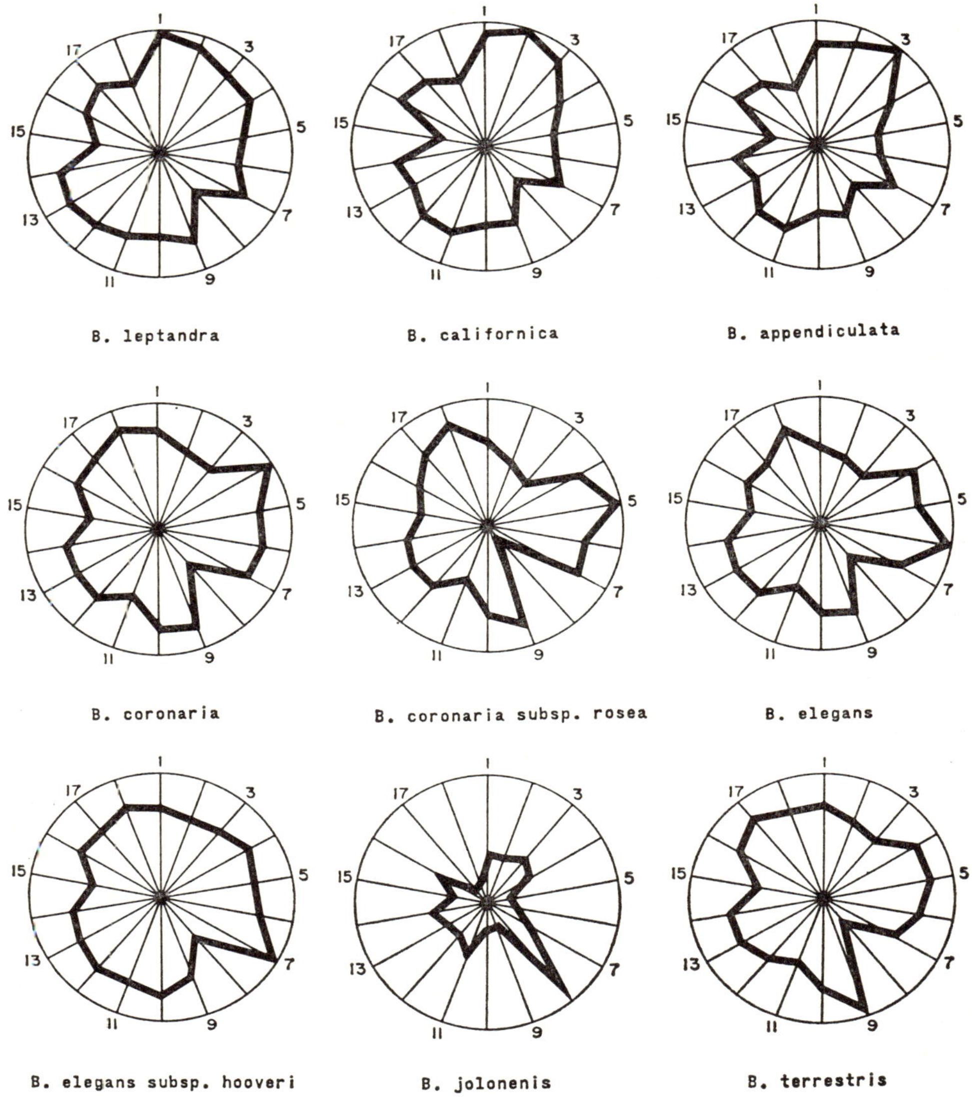

Fig. 3. Polygonal representations of paired affinity indices for each species of *Brodiaea*.

dices are plotted, the chemical similarity agrees with the species clusters (taxonomic sections) obtained using floral morphology.

Examples of this pattern are the trio *B. appendiculata*, *B. californica*, and *B. leptandra*, which have similar floral and chemical characters. *Brodiaea kinkiensis*, *B. filifolia*, and *B. orcuttii* represent another cluster of species having similar floral and chemical features. However, the affinity polygon for *B. jolonensis* indicates a low affinity with all other species. *Brodiaea jolonensis* has seven flavonoid compounds in contrast to the eleven or more in the other species of *Brodiaea*. Table 6 shows that *B. jolonensis* lacks many of the flavonoid compounds found

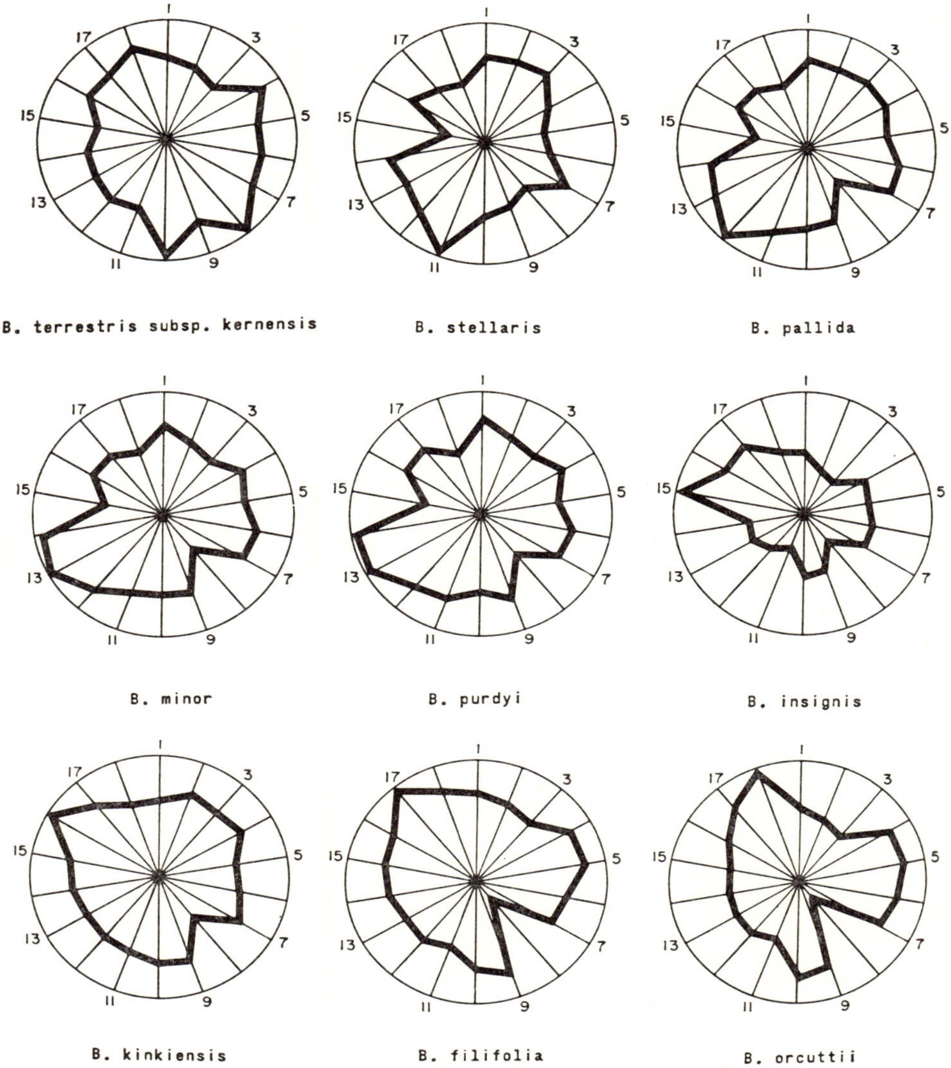

Fig. 3 Continued.

in all other species. Two possible explanations for this difference are that *B. jolonensis* was segregated from the rest of the genus very early in the evolutionary history of the genus or that it has recently changed to different biochemical pathways. If the most similar flavonoid compounds between all species represent the lower or beginning portions of the flavonoid biosynthetic pathway, *B. jolonensis* would not be representative of a more primitive species. That is, flavonoid compounds found in *B. jolonensis* are the result of a recently changed biochemical pathway. Identification of the flavonoid compounds would be an important step toward understanding why *B. jolonensis* is so different chemically from all other species of *Brodiaea*.

CHROMOSOME STUDIES

Meiotic Chromosomes

The chromosome numbers of only six of the eighteen taxa in *Brodiaea* were known prior to my study. Johansen (1932) reported chromosome counts for three

TABLE 7

Haploid Chromosome Numbers in Brodiaea

Species		Reference
Section Appendiculatae		
B. appendiculata	6	Niehaus, 1965
B. californica	6	Burbanck, 1941
B. leptandra	6	Niehaus, 1965
	12*	
Section Coronariae		
B. coronaria	6*	
	12	Niehaus, 1965
subsp. *rosea*	6	Niehaus, 1965
B. elegans	6*	
	8	Niehaus, 1965
	12*	
	16	Burbanck, 1941
subsp. *hooveri*	20*	
Section Terrestres		
B. terrestris	6*	
	18	Niehaus, 1965
subsp. *kernensis*	24*	
B. jolonensis	6	Niehaus, 1965
	18*	
Section Stellares		
B. stellaris	6	Johansen, 1932
B. pallida	6	Niehaus, 1965
B. minor	6	Burbanck, 1941
B. purdyi	12*	
	24*	
B. insignis	16*	
Section Filifoliae		
B. kinkiensis	16†	Niehaus, 1965
B. filifolia	12*	
B. orcuttii	12	Lenz, 1966

*Chromosome number not previously reported.
†Reported as *B. filifolia*.

species and Burbanck (1941) for four (table 7), but their counts do not entirely agree. For example, Johansen reported n = 5 for *B. californica,* while Burbanck reported n = 6. Some of my observations do not agree with those of either of these workers. The discrepancy is discussed below.

Meiotic chromosomes of all taxa in *Brodiaea* were observed. Flower buds with anthers at the appropriate stages of meiosis were fixed in Carnoy's solution (3 parts absolute ethyl alcohol: 1 part glacial acetic acid) and stored in a freezer until examined. The aceto-carmine smear technique was used to make microscope slides of pollen mother cells undergoing meiosis (Belling, 1926). More than one plant was examined in each collection; samples varied from two to ten plants, with three or four the usual number.

The meiotic chromosomes of *Brodiaea* are large and easy to observe even at low levels of magnification. Inadequate fixation of flower buds may result in clumping of chromosomes, which makes it difficult to obtain an accurate count. The time of day that buds are fixed determines the stage of meiosis observed. The best time for metaphase I was 10 P.M.; the later stages of meiosis occurred between noon and 2 P.M. The material collected in late afternoon or early evening (5–8 P.M.) was in prophase I or early metaphase I.

Chromosome counts for the genus *Brodiaea* are summarized in table 7. The species are arranged in five sections on the basis of similarities of floral morphology. Chromosome number and chromosome morphology fit the same general pattern of species clusters (sections of *Brodiaea*). A majority of the species have $n = 6$, which I consider to be the base number for the genus because of its presence in most of the taxa and because most polyploid numbers are multiples of 6. In the related genera the basic numbers for *Triteleia* are $x = 7$ and 8; for *Dichelostemma*, $x = 9$. One or more levels of polyploidy were found in many species: *B. coronaria* has $n = 6$ and 12; *B. elegans*, $n = 6$, 8, 12, 16, and 20. No meiotic irregularities or quadrivalents were observed in any polyploids. Autoploids of other angiospermous genera frequently form quadrivalents at the tetraploid and higher levels of polyploidy (Stebbins, 1950). The absence of quadrivalent formation in *Brodiaea* suggests that the polyploids are alloploids. However, the fact that the diploids and the teraploids of *Brodiaea* appear morphologically similar suggests that autoploidy rather than alloploidy is present. It has been suggested (Swanson, 1960; Ornduff, 1969) that autoploidy does not necessarily require quadrivalent formation.

Johansen's (1932) chromosome count for *B. californica* ($n = 5$) is incorrect so far as I can determine. His drawing of *B. californica* chromosomes shows one of the five metaphase I chromosomes as much larger than the others. The larger chromosome is probably due to clumping of two chromosomes; the same appearance was observed in one of my poorly fixed *B. californica* samples. Johansen (1932) reported $n = 21$ for *B. coronaria* from western Oregon. Burbanck (1941) reported $n = 21$ and $2n = 42$ for plants of *B. coronaria* from near Corvallis, Oregon. Collections from the Corvallis area and elsewhere in the Willamette Valley were found in my studies to have a chromosome number of $n = 20$ or $2n = 40$. The counts of $n = 21$ or $2n = 42$ obtained by Johansen and Burbanck were probably due to excessive pressure while making slides, which can result in fragmentation of chromosomes: I have found that it will fragment or separate *Brodiaea* chromosomes. Plants from the Corvallis area are here referred to *B. elegans* subsp. *hooveri* rather than to *B. coronaria* (Burbanck, 1941).

Burbanck (1941) reported counts of $n = 16$ and $2n = 32$ for *B. purdyi* from corms of uncertain identity sent to her by Carl Purdy. I have found $n = 12$ or 24 for all populations of *B. purdyi* examined in this study. In my opinion the Burbanck counts for *B. purdyi* should be disregarded.

The most widespread species of *Brodiaea* is *B. elegans*. Chromosome numbers of n = 6, 8, 12, and 20 have been observed. Most populations of *B. elegans* have n = 16, but those from the higher foothill region of the Sierra Nevada have n = 8. *Brodiaea elegans* subsp. *hooveri,* which is found only in Oregon, has n = 20. The change from n = 8 to n = 16 may be the result of autopolyploidy. The populations with n = 20 could have developed by autopolyploidy plus some faulty divisions or by hybridization of n = 8 × n = 12 entities.

TABLE 8

MEIOTIC CHROMOSOME COUNTS FROM SAMPLE POPULATION OF BRODIAEA ELEGANS "VAR. MUNDULA"

Haploid no.	Observations
6	Several bridges A-I, 10 percent of cells
6	M-I
6	M-I
6	M-I
6	M-I
6	M-I
6	M-I
6	M-I
8	M-II, A-II
8	M-I, A-I univalents
12	Several bridges A-II, 10 percent of cells
12	M-II
12	M-II
16	A-II
16	M-I, A-I, M-II, A-II
16	M-I

East of Sonora, Tuolumne County, California, contact has occurred between populations having n = 16 and others having n = 8. A peculiar form of *B. elegans,* called "var. *mundula* Jepson," restricted to a very local area on two sides of one small hill, is characterized by very short pedicels, 0.2 to 1.0 cm long versus 0.6 to 10.0 cm in *B. elegans* subsp. *elegans.* Variety *mundula* always occurs with *B. elegans* subsp. *elegans* and comprises approximately half of the population on the north side of the hill and 5 percent or less on the south side. Cytologically, *B. elegans* subsp. *elegans* was found to be n = 12 and 16, whereas "var. *mundula*" can be n = 6, 8, 12, or 16; all these chromosome counts are from the population on the north side of the hill (table 8). This is the only known occurrence of n = 6 and n = 12 for *B. elegans.* Plants of *B. elegans* from this population were examined for pedicel length, which varied from 0.2 cm to 10.0 cm. Since there is a transition in pedicel length without any sharp boundary, "var. *mundula*" is not accorded taxonomic recognition here.

On one slide of pollen mother cells of "var. *mundula*" approximately 25 cells at anaphase I had 12 chromosomes at each pole. In a single cell of the same preparation there were 8 chromosomes at one pole and 16 at the other. This misdivision suggests a possible mechanism for the derivation of n = 8 and n = 16 ploidy levels in *B. elegans.* Bud samples of "var. *mundula*" were obtained from widely separated plants several weeks before flowering began so the exact morphology for each of the plants reported in table 8 is not known. A detailed knowledge of the floral morphology of each plant would be useful in deciding whether length of pedicel is related to chromosome number or to some other factor such as multiple gene inheritance.

The existence of the "var. *mundula*" population with n = 6, 8, 12, 16 can be explained by chance hybridization between n = 16 and n = 8 plants. These cytotypes occur, respectively, from the "var. *mundula*" site (elevation 2,000 feet) to lower elevations, and from 2,000 feet to the upper limit of distribution of *B. elegans* proper. Many of the plants in this contact zone (table 8) appear to be cytologically unstable. The chromosome numbers n = 6 and 12 in the "var. *mundula*" population appear to be derived, for they are not known elsewhere for *B. elegans.* These might possibly represent an example of depolyploidization, vis., the reduction of tetraploid to a diploid (Kimber and Riley, 1963; Raven and Thompson, 1964; Ornduff, 1969).

Mitotic Chromosomes

Initial studies showed that mitotic chromosomes differ in size between species. The study was then expanded to all species in the hope that chromosome morphology could be used as an additional character to relate one species with another.

When *Brodiaea* corms break dormancy in the fall, rapid root elongation occurs. To obtain root tips for mitotic chromosome studies, corms were placed in a controlled environment chamber with an eight-hour day and a constant temperature of 40° F. After two days of growth the root tips were removed and fixed in Carnoy's solution. A majority of division figures were at metaphase two hours after the lights of the growth chamber turned on after sixteen hours of darkness. All root tips used in this study were obtained under these conditions.

The root tip was hydrolyzed in 15 ml of 70 percent ethyl alcohol to which several drops of 1N hydrochloric acid were added. After four or five minutes in the hydrolyzing solution the root tip was transferred to a drop of aceto-carmine and squashed. This procedure resulted in good slides in which all the chromosomes were well spread. The chromosome complement was then drawn with the aid of a Zeiss Abbe drawing apparatus. The drawn chromosome complement for each species is presented as an idiogram (fig. 4).

The idiograms show that there is a rather wide range of chromosome size in *Brodiaea.* The species with the longest individual chromosomes also have the largest chromosome volume (*B. jolonensis, B. appediculata,* and *B. californica*) and the converse is true of the smallest (*B. filifolia* and *B. orcuttii*). Primitive species have a larger total volume than more advanced species observed in *Crepis* (Babcock, 1947). Most chromosome complements are similar and most chromosomes in each complement are acrocentric.

The changes in idiogram can be correlated with progressive floral specialization. The best example is the series *B. kinkiensis—B. filifolia—B. orcuttii,* in which the staminodia range from partially reduced structures (*B. kinkiensis*) to completely obsolete (*B. orcuttii*). Chromosome size decreases in the same sequence, from *B. kinkiensis—B. filifolia—B. orcuttii.* Progressive reduction in chromosome size

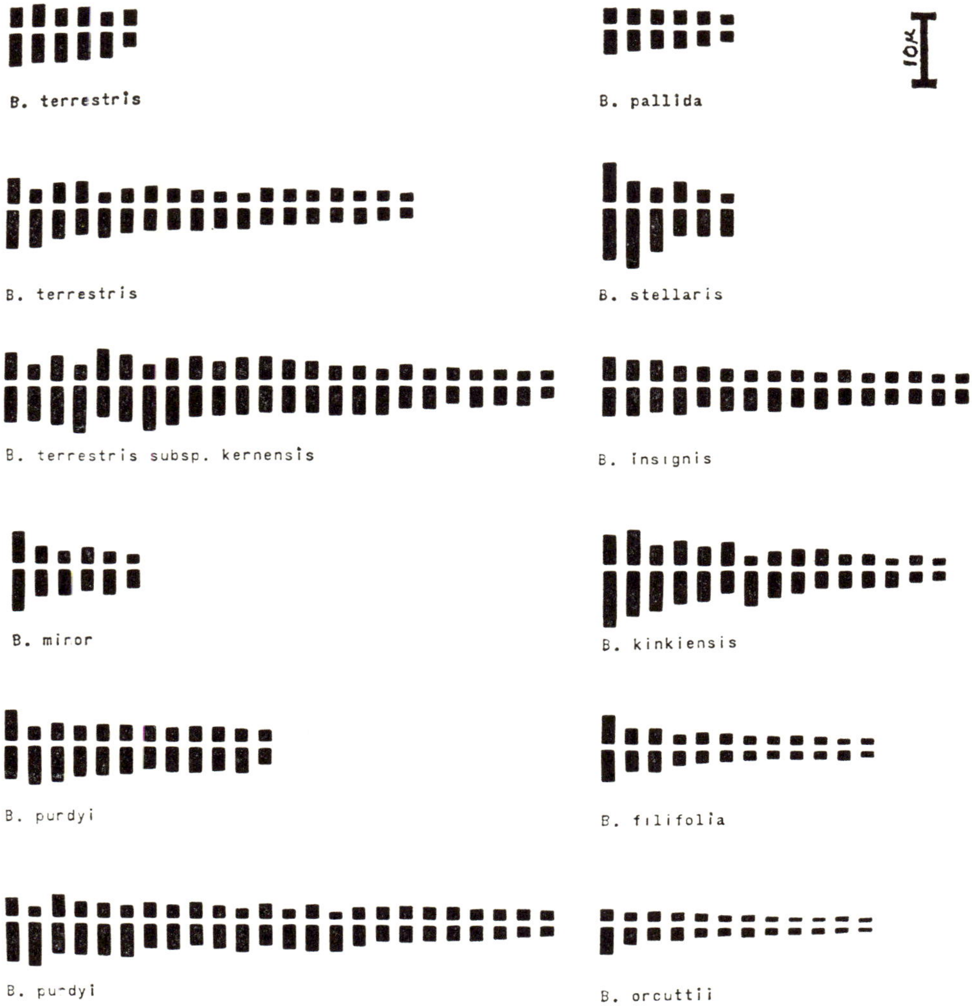

Fig. 4. Idiograms of *Brodiaea* chromosomes. Each idiogram is arranged by longest total length of each chromosome with a progressive decrease in size from left to right.

characterized other clusters of species also, and is discussed in the section on evolution.

When chromosome complements of polyploid series in a single taxon are compared, the general symmetry of each chromosome complement of a species and its subspecies (diploid, tetraploid, etc.) appears at first glance to be the same (e.g., *B. terrestris* subsp. *terrestris* and *B. terrestris* subsp. *kernensis*). Both species and subspecies ploidy complements are compared on the assumption that they are all autoploid derivatives within a species. However, a close comparison of individual chromosomes shows that an exact duplication of the chromosome complement of a diploid ancestor does not occur in the resultant tetraploids, hexaploids, and octoploids (*B. terrestris* subsp. *terrestris, B. terrestris* subsp. *kernensis, B.*

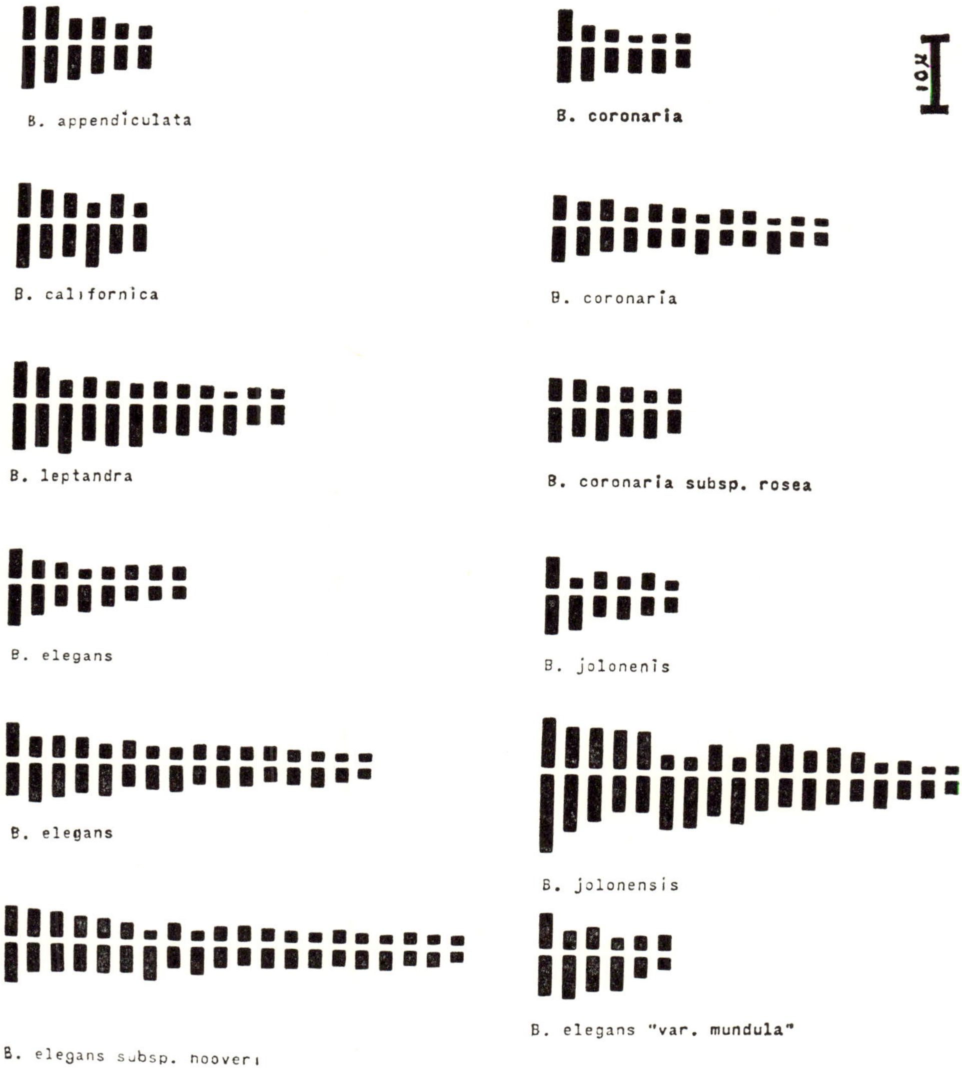

Fig. 4. Continued.

coronaria subsp. *coronaria, B. coronaria* subsp. *rosea*). Chromosomal symmetry between complements at different ploidy levels would be expected if the ploidy increase is due solely to autoploidy. Since the ploidy increase is within a given species, autoploidy is suspected. The lack of quadrivalents in meiosis plus the differences in size of individual mitotic chromosomes at different ploidy levels suggest that these are "old" autoploids. That is, the tetraploids, hexaploids, and octoploids have been in existence long enough to change by translocations, deletions, and so on. These differences apparently are enough to ensure that during meiosis only bivalents will occur in the higher ploidy levels.

Earlier it was suggested that depolyploidization may be occurring in *B. elegans.*

TABLE 9. PERCENTAGE OF SEED-SET OF ARTIFICIAL F 1's.*

	AP		CA		LEP		CO		RO		EL		JO		TE		KE		ST		PA		MI		PU		IN		KI		FI		OR
AP																																	
CA		13.3																															
LEP	9.0	5.0		50.0																													
CO		16.8				21.5																											
RO		38.7	2.0	40.0			72.5	33.1																									
EL		10.9	27.9			13.0	6.7	50.2																									
JO	57.8		5.6			8.5	58.4	36.9	51.6	49.4	2.8	6.3																					
TE	4.3	13.7			13.0		100	22.2		10.7		10.8																					
KE		2.7					62.5	17.3			24.7	15.8	55.2	6.4	34.2	8.6																	
ST					11.5		8.6	17.9	15.8		15.5					6.5	17.2																
PA		2.7	55.6	24.0				2.5	21.2	96.8	46.3	8.3	24.1	46.3	55.5		27.7	6.5		19.1													
MI	1.0		49.3		43.3	5.6			85.3		9.3	8.3	50.0		10.6		10.0				10.5												
PU	27.0		25.6		14.8	37.5	52.0	51.0			9.5	12.5	6.7		82.4	21.7	21.6	85.5	6.7	9.5	6.7												
IN		5.0						53.6		21.5				49.1		4.3		49.3				9.2				22.0							
KI	38.2		44.1		58.8		70.0	20.4	40.6	48.3	34.3		13.7	10.7	18.1	4.6	17.6		66.1	24.7	12.7	64.8	31.8	8.1	66.1	62.6	35.7						
FI			48.0		14.4		100		9.6	21.5	52.8	4.2	64.4	22.3	38.4	32.6	57.7	26.3		12.9	25.9	78.7		25.6		8.4		38.4	38.4	36.7			
OR	4.7	16.4	11.7		34.4		62.0	20.0	7.0	10.7	25.9	8.3	18.8	26.5	16.5	10.4	18.8		4.7	21.6	18.8	27.7		8.7	4.7	18.0		9.4	48.4	48.0	42.9	17.3	

* The number on the left side represents seed-set percentage on female plant as indicated on left side of table. The number on the right side represents seed-set percentage on female plant as indicated on top side of table. The specific name of each species is represented by the first two letters of its name.

The chromosome complement (2n = 12) of "var. *mundula*" could theoretically be derived by crossing 2m = 16 and 2n = 32 strains of *B. elegans* to form a hybrid with 2n = 24. It would then be necessary to reduce this hybrid to 2n = 12; the exact mechanism for such a reduction is unknown. However, the suspected hybrids of the two strains (see table 8) are markedly unstable, and it is possible that the "var. *mundula*" karyotype could be made up of individual chromosomes from both strains. The "var. *mundula*" mitotic complement (fig. 4) required considerable averaging to arrive at a measurement for each pair of chromosomes in the idiogram. Only a few "var. *mundula*" complements of mitotic chromosomes have been compared with those of *B. elegans*, and additional comparisons would help in assessing the stability of "var. *mundula*."

HYBRIDIZATION STUDIES

Artificial hybrids between species can be useful in determining how closely they are related (Stebbins, 1950; Lewis and Epling, 1959; Thompson, 1960). Hybrids can provide information from (1) crossability, (2) seedling vigor, (3) chromosome pairing, and (4) pollen fertility. The data may then be used to determine whether there are genetic barriers between species and how these affect the possibility of interspecific hybridization in nature. Hybrids must survive chromosomal lethalities and ecological competition. Depending upon whether they do so, it may be possible to determine whether closely similar morphological species are the ends of separate phylogenetic lines or are derived one from another. With these factors in mind, a program of artificial hybridization between each pair of species was undertaken during the springs of 1966 and 1967.

All plants used in this study were grown inside a greenhouse in which the ends are screened against birds but open to local insects. During the flowering period insects were rarely observed to visit *Brodiaea*. Some emasculated flowers were left unpollinated, and these were not pollinated even though left for a week or more. Before attempting hybridization, the flower of the prospective seed parent was emasculated with forceps as soon as it opened. It was allowed to remain unpollinated for several days to ensure that the stigma was receptive. The flower was then pollinated with pollen from a newly opened flower. Ten flowers of a species from one locality were used for each combination. Crosses were made between plants originating in different populations of each taxon to determine whether intraspecific sterility was present. In order to detect self-compatibility, self-pollinations were made within certain populations. Flowers in which pollen-tube growth apparently occurred closed within a day after pollination and the ovary began to swell. Seed was collected just as the capsules began to open. The number of seeds in each capsule was recorded and the percentage of seed-set was calculated for each cross that produced seed. The percentage of hybrid seed-set is compared with average seed-set of the female parent in intraspecific crosses within a specific population of each species. This is recorded in table 9 for interspecific hybrids of *Brodiaea*.

No self-pollinations produced any seed. This indicates that all *Brodiaea* species are self-incompatible. However, no information was obtained concerning the type of self-incompatibility system in the genus. Table 9 shows that most of the crosses between species produce a low percentage of seed-set. In a few crosses a relatively high level of seed-set was obtained. Seed-set used as a measure of crossability shows that many *Brodiaea* plants have an isolation barrier in the form of little or no seed-set after interspecific crosses. Seed-set appears to show no consistent pattern of crossability even in species comprising a section of the genus. *Brodiaea* plants grown from seed require two to three years or more before flowering occurs. Once seeds have germinated, all the hybrid seedlings appeared to be vigorous. Only one cross had produced flowers at the time this paper was written, a hybrid between *B. terrestris* and *B. coronaria,* which had a 100 percent seed-set, a figure rarely reached in any of the other crosses. The progeny of this cross had flowers that were morphologically intermediate between those of the two parents. Pollen fertility of each hybrid plant was obtained: the majority were completely sterile, but one plant did show 92 percent fertility. Interpretation of the hybridization studies must remain incomplete until more information is available. Interfertility may be interpreted in terms of a single criterion such as seed-set, but it has greater meaning when (1) seed-set, (2) seedling vigor, (3) chromosome pairing, and (4) pollen fertility are considered together.

ECOLOGY

General Observations

The growing cycle of *Brodiaea* is dependent on the climatic regime of the western United States and more specifically that found west of the Sierra Nevada–Cascade axis. The corms, which are dormant during the summer, begin growing after the first significant fall rains, when the soil surrounding the corm is moistened to or nearly to field capacity. The roots elongate rapidly; the leaves also begin to elongate and continue to do so slowly throughout the winter. The leaves appear to be tolerant of light frost, but a hard freeze can injure them. The upper limit of distribution for *B. elegans* is at or near the winter snow line. All other species of *Brodiaea* are found at lower and presumably warmer elevations.

The species found at the lowest elevations are the first to flower in mid-April (*B. minor, B. appendiculata*), with a progression toward the later-flowering species found at higher elevations (*B. purdyi, B. elegans*). It is well known that the flowering season in California can be early or late depending on the amount of moisture and the temperature early in the spring. Thus the flowering date of *Brodiaea* is affected by the spring weather in any particular year. A guide to the flowering of any species of *Brodiaea* can be predicted on the basis of the condition of mid-spring herbs and the introduced grass, *Avena.* When daytime temperatures have reached daily highs in the 80's and 90's and many of the herbs and grasses have died and turned brown, the *Brodiaea* species begin flowering. Thus the members of *Brodiaea* can be thought of as early summer bloomers.

At the time of flowering the leaves of *Brodiaea* are dead or nearly so, and the corms that will start the next generation in the fall are mature. The flowering period usually lasts from two to three weeks. Development of capsule and seeds require four to eight weeks before growth ceases. By the time the capsule is mature the scape and pedicel are dead. *Brodiaea* plants grow only in full sun and, generally speaking, in clay soil. *Brodiaea* is frequently associated with soils that are highly saturated with water during the winter. Some species, such as *B. coronaria, M. minor, and B. orcutti,* are found along the margins of vernal pools or intermittent streams; others, such as *B. elegans,* and *B. terrestris,* are seemingly less water-dependent.

Pollination

Since plants of *Brodiaea* are self-incompatible, pollination between individuals must occur for seeds to develop. In areas where species are sympatric, few interspecific hybrids are known; this suggests that different pollinators might be involved. Hence the method of pollination was examined. Because of time limitations, an extensive study of pollination biology was not carried out. However, insects of the orders Diptera (Family Bombyliidae—Bee Flies), Lepidoptera (Family Hesperiinae—*Hesperia ruricola* Bdw.—Skipper Butterfly), Coleoptera (Family Mordellidae—Tumbling Flower Beetle), and Hymenoptera (Halictidae —Sweat Bees) were frequently observed visiting *Brodiaea* flowers.

Insects that frequent *Brodiaea* were examined to see how much pollen is carried to different plants. Although Diptera and Lepidoptera often visit *Brodiaea* for nectar, they do not carry its pollen. These observations were confirmed by study of insects visiting the flowers. The long tongues of Diptera and Lepidoptera obtain nectar from the flowers, but their bodies do not come in contact with the anthers. No *Brodiaea* pollen was found on the bodies or tongues of these insects. The Coleoptera and Hymenoptera that visit *Brodiaea* collect pollen for food. They frequently crawl into the flower between anthers and stigma and a few minutes later fly to another flower. Considerable pollen is entrapped by the hairs of their bodies. Thus pollen transfer is possible from one plant to another.

Although beetles are scarce when a particular species begins to flower, by peak bloom ten or more are frequently seen on a single flower. Hymenoptera usually visit one flower for less than a minute before flying on to the next. They also were active in visiting flowers during high winds, when the beetles were unable to fly.

Distribution of Species in Relation to Soil Types

Field observations indicate that each species of *Brodiaea* occurs on only one or a few soil series. Early in my field work I noted that *B. minor* grew only on a red clay soil of the Corning soil series and *B. appendiculata* on an adjoining red gravelly clay of the Redding soil series. This suggested the possibility that some or all *Brodiaea* species were limited to specific soils. Because adjoining soil series support plants of different but closely related species, it is hypothesized that as neighboring soils became available, derived forms had a new habitat for colonization and evolution. Soil types were determined in the field and known localities for a given species were compared by means of Soil Survey Maps (U.S.D.A.), which were sometimes used to predict the site of a species before visiting the area. This method of determining localities was always reliable.

Since the Department of Agriculture surveys are primarily of land potentially useful for agriculture, only partial information is available for nonagricultural areas of soils in which most members of *Brodiaea* grow. The distribution of each species of *Brodiaea* was compared with the available Soil Survey Maps. Table 10 shows the soil series in which each species is presently known to occur. The soil series of California are grouped in five categories on the basis of land form, parent material, and mode of development. The five groups are: (1) soils of recent and young alluvial fans; (2) soils of recent flood plains; (3) soils of alluvial basins; (4) soils of old terraces and old alluvial fans; (5) soils of the hills and mountains.

TABLE 10

SOILS IN WHICH BRODIAEA GROW

Species	Soil series	Type
B. appendiculata	Corning	Old Terrace
	Keefer	
B. californica	Redding	Old Terrace
B. leptandra	Altamount	Residual
B. coronaria	Tuscan	Recent Alluvial
	Wyman	
subsp. *rosea*		Serpentine
B. elegans	Bale, Butte	Old Alluvial
	Keefer	Residuals
	Altamount	
subsp. *hooveri*	no information	
B. terrestris	Altamount	Residuals
	McClusky	Old Terrace
subsp. *kernensis*	Altamount	Residual
	Cuyama	Alluvial
	Delano	Alluvial
B. jolonensis	Redding	Old Terrace
	Pleasanton	Old Alluvial
B. stellaris		Serpentine
B. pallida		Serpentine
B. minor	Corning	Old Terrace
B. purdyi		Residuals and Serpentine
B. insignis		Residual
B. kinkiensis		Old Alluvial
B. filifolia		Recent Alluvial
B. orcuttii	Redding	Old Terrace

It is significant that *B. appendiculata, B. californica,* and *B. minor* are found on soils of old terraces and old alluvial fans along the eastern margin of the Central Valley of California, and that *B. jolonensis, B. terrestris,* and *B. orcuttii* occur in similar soils along the coast of southern California. These soils are all characterized by gravelly clay which may or may not have a hardpan. *Brodiaea terrestris* subsp. *kernensis, B. leptandra, B. purdyi,* and *B. insignis* are found on residual soils adjacent to the old alluvial but at slightly higher elevations in the Sierra Nevada foothills or in the Coast Ranges. The latter soils are clay-loam developed from basic igneous rocks in the area. A third type of soil, which supports *B. stellaris, B. pallida, B. purdyi,* and *B. coronaria* subsp. *rosea,* consists of clays derived from serpentine rock.

Information on the age of these soils is meager, since no methods have yet been devised to date them accurately (Rodney Arkley, pers. com.). The only ages known are those of old terrace and alluvial soils, which are thought to have been deposited in early Pleistocene time (Turner, 1894).

On the basis of restriction to soil type, *Brodiaea* species show the same evolutionary pattern suggested by their floral morphology and chromosome numbers. Species which grow on old terraces or old alluvials can be considered as the least advanced; those found on residual soils are more advanced evolutionally. The most advanced are the species growing on serpentine residual soils. Only *B. orcuttii* is inconsistent with this general pattern: on the basis of reduced flower size, loss of staminodia, and reduced chromosome size it is the most advanced species of *Brodiaea*. It grows, however, on an old alluvial terrace of Pleistocene age in the San Diego area. This is in marked contrast to other advanced species, all of which grow only on more recent residual soils. However, the present habitat of *B. orcuttii* was submerged until the end of the Pleistocene (Jahns, 1954), and thus *B. orcuttii* is apparently a post-Pleistocene arrival.

DISCUSSION

General Discussion

In prior taxonomic work on the genus *Brodiaea* little attempt has been made to explain relationships among the species (Jepson, 1925; Hoover, 1939*b;* Munz, 1959). Hoover (1939*a,* 1939*b*) employed morphological features in his investigation of relationships within *Brodiaea.* He concluded that four distinct sections or species clusters are definable in *Brodiaea.* He used perianth-tube shape, staminodia position, and stamen shape as his criteria for taxonomic sections. I believe that three of his sections (Appendiculatae, Stellares, Filifoliae) are valid and that the fourth, Coronariae, should be divided into two: Coronariae and Terrestres. I found that *B. terrestris* and *B. jolonensis* have a distinct series of floral characters which differ from the other members of Coronariae and thus should be placed in their own section.

Hoover's (1939*b*) revision was an important step forward in our taxonomic knowledge of the genus. He published several new species and solved a number of taxonomic problems. For instance, Hoover concluded that *B. coronaria* as interpreted by Jepson consisted of two distinct species, *B. coronaria* and *B. elegans. Brodiaea coronaria* (Salisb.) Engler is common in Washington but of infrequent occurrence in California, whereas *B. elegans* Hoover is common in California and reaches its northern limit in Oregon. Hoover, whose boyhood home was near Oakdale, California, did extensive field work in the nearby foothills of the Sierra Nevada. He thus became aware of the existence of four undescribed species of *Brodiaea* in this region. His field work was less intensive in other areas of *Brodiaea* distribution, for which he had to rely on herbarium specimens. Even so, he was aware that *B. elegans* in Oregon, for example, was slightly different from California material, but without fresh flowers he was unable to do more than report this variation. This Oregon variant is here distinguished as *B. elegans* subsp. *hooveri,* in recognition of Hoover's important contributions toward a knowledge of the genus.

Hoover recognized ten species and six varieties of *Brodiaea* in his treatment

(1939*b*). On the basis of my examination of a much wider range of characters, I have concluded that fifteen species and three subspecies deserve recognition. Two of the additional species were discovered during the investigation. Study of the morphology of each taxon yielded many characters which had not been used previously. For example, some species can be identified by leaf habit and number of leaves. Floral morphology has contributed most of the characters that differentiate the species. To supplement external morphological characters, I studied vascular anatomy, pollen, chromatographic patterns of flavonoids, chromosome cytology, and ecological features, and conducted an artificial hybridization program. Each of these aspects of the study yielded additional characters that are useful in distinguishing species. When all characters of each species are considered together, each appears as a discrete unit.

Evolution in Brodiaea

A phylogenetic scheme for *Brodiaea* has been constructed on the basis of data from studies of the morphology, chemistry, cytology, ecology, and distribution of each species. As figure 5 indicates, some species are derived directly from others, and other species in the same cluster (taxonomic section) are related to each other but less directly so.

Section Appendiculatae is a natural group characterized by large flowers in which the perianth tube is cylindric and rounded at base and becomes brittle and splits in fruit. Perianth segments are ascending and the elongated staminodia are linear. This section comprises three species; their floral morphology suggests that *B. appendiculata, B. californica,* and *B. leptandra* are derived successively in that order. A forked appendage at the apex of the stamen filament characterizes *B. appendiculata. Brodiaea californica* is similar in most features to *B. appendiculata* except that this forked appendage is reduced to a small remnant or is lacking in some plants in a population. The perianth tube of *B. californica* is shorter than that in *B. appendiculata. Brodiaea leptandra* resembles *B. californica* but lacks a filament appendage and all floral parts are narrower. Corm coats of *B. appendiculata* consist of thick fibers; *B. californica* and *B. leptandra* have smaller fibers. Lateral and median vascular traces of the inner perianth segments of *B. appendiculata* and *B. californica* are single-stranded; those of *B. leptandra* are double-stranded. Flavonoid chemistry suggests the same relationship of species (fig. 3). *Brodiaea appendiculata* and *B. californica* both have $n = 6$. Although one population of *B. leptandra* has $n = 6$, $n = 12$ is otherwise known for this species. Chromosome morphology is slightly different between *B. appendiculata* and *B. californica.* In *B. leptandra* the chromosome complement is usually tetraploid and the chromosomes are smaller than those of the other two species.

Brodiaea appendiculata grows in old terrace alluvial soil along the lowest foothills of the Sierra Nevada from near Madera to Sacramento, and flowers from mid-April to mid-May. *Brodiaea californica* occupies another type of old terrace soil from Nevada County to Shasta County at slightly higher elevations, and in some areas is found also on serpentine soil; it flowers from mid-May to late June. *Brodiaea leptandra* is restricted to a few areas in the North Coast Ranges and occurs in a residual soil; it flowers in June. When all the characters are con-

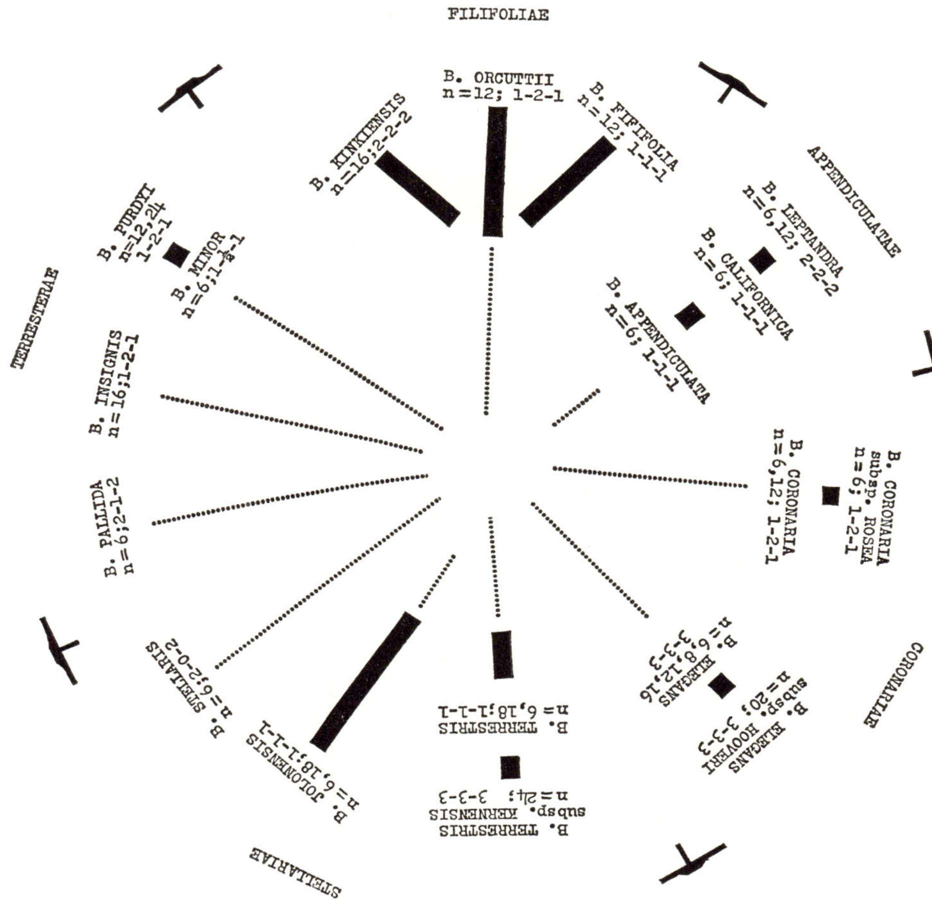

Fig. 5. Phylogenetic scheme for *Brodiaea* is shown as a series of species whose ancestors are either probable (solid lines) or unknown (dotted lines). Distance from each other or from the center does not imply phyletic distance. Numbers below the name of the taxon represent chromosome numbers and vasculature of the inner perianth segments, respectively.

sidered together, a strong trend from *B. appendiculata* to *B. californica* and then to *B. leptandra* is indicated.

Section Coronariae is characterized by a perianth tube of tough texture which does not split in fruit. The perianth segments are recurved and the staminodia are strongly involuted or flattened. This section consists of two species, *B. coronaria* and *B. elegans,* of unknown derivation. *Brodiaea coronaria* has one highly localized subspecies *rosea;* morphologically *B. coronaria* subsp. *coronaria* and *B. coronaria* subsp. *rosea* are rather similar. Corm coats of *B. coronaria* subsp. *coronaria* are composed of heavy fibers; those of *B. coronaria* subsp. *rosea* are less fibrous. *Brodiaea coronaria* subsp. *coronaria* has a haploid chromosome complement of n = 6 in northern California and southern Oregon, and n = 12 from Oregon to British Columbia and also in central California. This fact suggests that the center of origin was in northern California or southern Oregon, that *B. coronaria* has migrated both northward and southward, and that autotetraploids may have arisen independently. *Brodiaea coronaria* subsp. *rosea* is a serpentine endemic that occurs in two places in the inner North Coast Ranges of California.

Brodiaea elegans subsp. *elegans* is the most widespread taxon of the genus; subsp. *hooveri* occurs at the northern end of its distribution. Along the upper foothills of the Sierra Nevada *B. elegans* subsp. *elegans* has n = 8, but elsewhere it has n = 16; subsp. *hooveri* has n = 20. At a contact zone between n = 8 and n = 16 of *B elegans* subsp. *elegans* near Sonora, California, hybridization has apparently resulted in n = 6 and n = 12 plants by misdivisions in the hybrid. This population is presently unstable cytologically, but it seems to afford an example of reduction in chromosome number. That is, from a triploid hybrid n = 6 and n = 12 plants were found to be more stable than a n = 8 hybrid plant which had only univalents.

Section Terrestres is characterized by smaller flowers and short, slightly involute staminodia with a partial hood along their apices. The two species of section Terrestres are closely related. *Brodiaea terrestris* subsp. *terrestris* grows on alluvial terraces along the Pacific Coast from Coos Bay, Oregon, to San Simeon in central California. At its southern limit it extends inland to near Paso Robles, where subsp. *kernensis* approaches it. *Brodiaea terrestris* subsp. *terrestris* has n = 18 throughout most of its distribution, with only one exceptional population of n = 6; *B. terrestris* subsp. *kernensis* is an octoploid with n = 24. The two subspecies are very similar morphologically except that in subsp. *kernensis* the vascular traces of the inner perianth segment are divided into triple instead of single strands. The larger flower of subsp. *kernensis* is probably related to its polyploidy. *Brodiaea jolonensis* is closely related to *B. terrestris* and apparently arose from the same ancestor. It has nearly the same floral morphology as *B. terrestris* except that the anthers lack a dentate appendage; however, the flavonoid chemistry of the two species is quite different. *Brodiaea jolonensis* and *B. terrestris* may have originated in the central coastal region of California, with *B. terrestris* occupying the region to the north and *B. jolonensis* that to the south.

Section Stellares is characterized by small flowers with the perianth segments rotately spreading and short, broad staminodia. *Brodiaea stellaris* is a serpentine endemic of the North Coast Ranges of California. It apparently has no close relatives. Similarly, *B. pallida* is a serpentine endemic known from only one location in the central Sierra Nevada foothills. Farther to the south, *B. insignis* is restricted to several localities in the Kaweah and Tule river drainages.

Brodiaea minor, the most widespread species in the section, is the putative ancestor of *B. purdyi,* and the two have similar floral morphology. In *B. minor* the inner perianth vascular traces have single lateral strands, the median strand is single at its base but doubled in the upper end of the perianth segment. It grows on old terrace soil from Merced to Red Bluff and flowers from mid-April to mid-May. *Brodiaea purdyi* has n = 12 in the southern range of its distribution and n = 24 in its northern range. It is found mainly on serpentine soil, less frequently on volcanic soils, and flowers in June.

Section Filifoliae is characterized by small flowers with a thin perianth tube which splits easily in fruit; staminodia are reduced or lacking in one species. *Brodiaea filifolia* and *B. kinkiensis* are closely related and may have been derived from the same ancestor. Staminodia are somewhat reduced in *B. kinkiensis* and even more so in *B. filifolia.* The staminodial vascular traces of *B. kinkiensis* are reduced in the upper end of the staminodia; in *B. filifolia* they are reduced throughout to a very thin thread. *Brodiaea filifolia* has n = 12 and *B. kinkiensis* n = 16. The origin of the chromosome number of *B. kinkiensis* is unknown. The mitotic chromosomes are strongly reduced in size compared with those found in other members of *Brodiaea*. *Brodiaea orcuttii*, with small flowers, no staminodia, polyploidy (2n = 24), and the most reduced chromosome size, is believed to be the most advanced species of the genus. It has many similarities with *B. filifolia* but its entirely different stamen characters indicate that *B. orcuttii* is not derived from *B. filifolia*.

Both *B. orcuttii* and *B. kinkiensis* grow on old terrace soils which were previously submerged by the sea but recently became available. *Brodiaea filifolia* also occurs on a recent alluvial soil. If the species of section Filifoliae are recently evolved members of *Brodiaea*, the availability of new habitats has allowed the colonization and evolution of *B. filifolia*, *B. kinkiensis*, and *B. orcuttii*. Section Filifoliae contains the three most advanced species of the genus, judging by their small flowers, polyploidy, and the reduced size of their chromosomes—a trend of specialization found by Babcock (1947) in *Crepis* as well as in numerous other genera (Stebbins, 1950).

The interrelationships of the sections are not clear. On the basis of geographic distribution, polyploidy, and morphological specialization, *Brodiaea* can be divided into different categories of endemics. Stebbins and Major (1965) have examined endemism and speciation in California to determine what floristic and ecological conditions promote the persistence of relict species and the origin of new species. The classification system for endemics on the basis of cytological data proposed by Favarger and Contandriopoulous (1961) was applied by Stebbins and Major to California plants. Four major kinds of endemics were proposed; paleoendemics, schizoendemics, patroendemics, and apoendemics.

Paleoendemics are taxonomically isolated ancient taxa that are possibly on the way to extinction, and are often ecologically specialized. This category has been applied to such genera as *Sequoia*, *Dendromecon*, and *Lynothamnus* and thus does not seem to pertain to *Brodiaea*.

Schizoendemics are the result of gradual speciation or more or less simultaneous divergence from a known parent. Many of the *Brodiaea* can be put in this category. *Brodiaea terrestris* and *B. jolonensis* are closely related and appear to have arisen from a common ancestor formerly present in central coastal California. *Brodiaea filifolia*, *B. kinkiensis*, and *B. orcuttii* seem to have diverged from the same or closely related ancestors; all three grow on recently available habitats in southern California. Although the ancestor of *B. insignis* is unknown, this taxon should be considered a schizoendemic, as will be explained below. The placement of *B. pallida* and *B. stellaris* in this classification is problematic. Both grow on serpentine soil in a specialized habitat (Kruckeberg, 1954), and have some morphological features in common, but there is no information on their ancestry. *Brodiaea coronaria* subsp. *rosea* is nearly identical morphologically with the typical subspecies, but has evolved as a serpentine endemic. *Brodiaea insignis*, *B. pallida*, and *B. stellaris* are members of the same species cluster. Their immediate ancestors are not known, but they do appear to be related and perhaps can be classified as schizoendemics.

Patroendemics are diploids which have given rise to polyploids. *Brodiaea minor* is an example of a diploid which has probably given rise to the polyploid *B. purdyi*. Similarly, the diploid *B. appendiculata* may have given rise to *B. californica,* another diploid, which in turn is the putative ancestor of teraploid *B. leptandra*.

Apoendemics are the polyploid derivatives of patroendemics; *B. purdyi* and *B. leptandra* are believed to be examples of these.

Thus the genus *Brodiaea* seems to contain the end points of several lines of evolution. Most of the early ancestors, including the connecting links between the different sections, are probably now extinct. Various mechanisms of evolution already known for other California genera are at work also in *Brodiaea*. Some floral structures have become increasingly complex; these include vasculature of corm coat and inner perianth segments. Vascular traces in the inner perianth segments have a pattern of increasing in number within each section of the genus (fig. 5). Whether increased vasculature is an advanced or a primitive condition is not known, but the vascular-bundle number does follow other trends in *Brodiaea*. Structures such as staminodia show reduction in size and vasculature until, in *B. orcuttii,* the staminodia completely disappear. The size of flowers in section Filifoliae is much reduced in comparison with that in the more primitive section Appendiculatae. Similar reductions in size and complexity of structures have been noted in *Crepis* (Babcock, 1947), in many other genera (Stebbins, 1950), and in the Polemoniaceae (Grant, 1959).

Many of the more specialized of the *Brodiaea* exhibit increasing polyploidy. These polyploids are frequently found on either more xeric or more recent habitats, in a manner similar to that described for species of *Clarkia* (Lewis, 1953). Along with increase in chromosome number there have been several instances of possible reduction, or depolyploidization, perhaps made possible by the presence of many duplicated gene loci (Stebbins, 1966). Reduction of chromosomal volume in the form of smaller chromosome complements has occurred in *Brodiaea* also, and is especially pronounced in *B. orcuttii,* the most advanced species. The reason for the trend toward smaller chromosomes in the more advanced members of a genus has never been explained satisfactorily (Stebbins, 1966), but it seems to have occurred also in *Crepis* (Babcock, 1947) and in the Polemoniaceae (Grant, 1959).

Edaphic climatic factors are important in the evolution of California genera (Mason, 1946*a,* 1946*b;* Kruckeberg, 1957). The more primitive *Brodiaea* species are found on old alluvial soil and the more advanced on serpentine soil.

The combination of morphological, chromosomal, and ecological specializations has resulted in isolating barriers that promoted the formation of the species complexes known at the present time.

TAXONOMY

Generic Concept

The first description of any member of what is now considered the genus *Brodiaea* appeared in 1808 in the proposal of the genus *Hookera* Salisb., based upon plants collected by Archibald Menzies near Fort Vancouver, Washington.

Smith, in the same year, but some two weeks after Salisbury's publication, proposed the generic name *Brodiaea* on the basis of a duplicate set of the Menzies collection. However, *Brodiaea* was not officially established until it was published in the Transactions of the Royal Horticulture Society in 1811. Smith had used *Hookeria* previously to designate a genus of moss, a prior use of the name *Hookera* which apparently was unknown to Salisbury at the time he proposed it for the Menzies collection. A bitter controversy ensued between Salisbury and Smith as to whose generic name was correct for *Brodiaea.* The problem was eventually resolved at the second International Botanical Congress, where *Brodiaea* was conserved over *Hookera* (Intern. Rules Bot. Nomen., 1906).

TABLE 11

GENERIC CHARACTERS OF BRODIAEA AND ALLIED GENERA

Character	*Brodiaea*	*Dichelostemma*	*Triteleia*
Corm-coat color	Dark brown	Brown	Straw
Abaxial leaf surface in x.s.	Rounded	Keeled	Keeled
Stomata	Sunken	Level	Level
Guard cells	Two ledges	One ledge	One ledge
Perianth base	Rounded	Obtuse	Attenuate
Stamen number	Three	Three (six in one sp.)	Six
Anthers	Basifixed	Basifixed	Versatile
Staminodia	Three	None	None
Ovary attachment	Basal	Basal	Stipitate
Stigma	Three winged	Three lobed	Not lobed
Seed surface	Longitudinal striations	Rectangularly raised plates	Hexagonally raised plates
Chromosome base number	x-6	x-9	x-7, 8
Habitat	Full sun Clay soils	Semi-shade Loams	Semi-shade Loams

Interpretation of the generic limits of *Brodiaea* became an increasing problem as additional species were discovered. At one time or another *Brodiaea* has been extended to include all amaryllids referred to *Hookera* Salisb., as well as other species now placed in *Dichelostemma, Triteleia,* and *Bloomeria.* Two schools of thought have emerged concerning the proper generic limits of *Brodiaea.* One school would prefer a single large genus comprising the three subgenera *Brodiaea, Dichelostemma,* and *Triteleia* (Jepson, 1925; Munz, 1959). The most recent treatment (Munz, 1959) places all three groups in *Brodiaea,* and includes some South American plants in the genus. The other school would elevate each subgenus to the generic level (Greene, 1886; Hoover, 1939*a*).

Keator, who recently revised *Dichelostemma* (1967), and I agree that the recognition of three separate genera is appropriate. A comparison of various distinguishing characters of these genera is presented in table 11 and figure 1. From this table it is obvious that the three genera are closely related but also that they are separated by many differences. Morphology of the leaf is sufficiently different to separate the three genera in the vegetative state. Flower shape and the arrangement and shape of stamens, ovary, and stigma are also distinct for each genus.

Chromosome numbers have now been determined for nearly all species of the three genera. Each genus has its own distinct series of chromosome numbers with a different base number: *Brodiaea* $x = 6$, *Triteleia* $x = 7$, 8, *Dichelostemma* $x = 9$.

The only characters which Munz gives to support the retention of a single inclusive genus are the common possession of a united perianth tube and the presence of corms. But on the basis of these shared characters related genera such as *Triteleiopsis* and *Androstephium* would also have to be placed in *Brodiaea*. The South American species referred to *Brodiaea* by Munz have a bulb rather than a corm and their leaves have an alliaceous odor not found in the North American tribe Brodiaeinae (Traub and Moldenke, 1955).

In my opinion the genus *Brodiaea* should be restricted to a small group of morphologically, cytologically, and ecologically related species found principally in California, with three species extending to Oregon and Washington.

Brodiaea Smith, Trans. Linn. Soc. London 10:2. 1811. *nom. conserv.*

Perennial herbs with basal leaves and an umbelliferous scape. Corm subglobose, with a dark-brown fibrous coat. Leaves linear, abaxially rounded and adaxially concave (or crescent-shaped in transection). Scape slender, rigid, terete. Inflorescence an umbel; flowers on individual pedicels subtended by scarious bracts. Pedicels articulated at base. Perianth of six segments, the lower portions united into a tube, the upper portion of the segments free, the outer three segments narrower than the inner. Stamens three, opposite the inner perianth segments, alternating with three staminodia (these absent in one species) opposite the outer perianth segments; stamens erect and approximate to the style, the filaments adnate (prominent on inner side of perianth tube), the anthers basifixed. Stigma three-parted, its lobes spreading and recurved, papillae on upper surface. Fruit a loculicidal capsule with two rows of seeds in each locule; seeds black with longitudinal surface striations. Chromosome numbers $n = 6$, 8, 12, 16, 18, 20, or 24.

Type species: Brodiaea grandiflora Smith, Trans. Linn. Soc. London 10:2. 1811. *Hookera* Salisb., Parad. Lond. pl. 98. 1808.

KEY TO SPECIES

A. Staminodia obtuse, truncate or notched.
  B. Flowers 24–38 mm long.
    C. Staminodia narrowly linear and slightly involute; perianth tube splitting in fruit.—Section Appendiculatae.
      D. Filaments with two filiform appendages. .... 1. *B. appendiculata*
      DD. Filaments without appendages (sometimes rudimentary in *B. californica*).
        E. Perianth segments 6–10 mm wide; ovary 7–11 mm long ...... 2. *B. californica*
        EE. Perianth segments 4–7 mm wide; ovary 5–7 mm long. .... 3. *B. leptandra*
    CC. Staminodia broad, either strongly involute or plane; perianth tube remaining intact in fruit.—Section Coronariae.
      D. Perianth tube ovoid to narrowly campanulate; staminodia involute, directed inward toward stamens. .... 4. *B. coronaria*
      DD. Perianth tube funnelform; staminodia erect and plane, distant from stamens. 5. *B. elegans*
  BB. Flowers 14–24 mm long.
    C. Staminodia leaning inward, margin semi-involute, apex broad and slightly hooded; perianth segments campanulate.—Section Terrestres.
      D. Scape 0.5–5.0 cm high; anther connective dentate; ovary green. .... 6. *B. terrestris*
      DD. Scape 6–15 cm high; anther connective entire; ovary light purple. . . 7. *B. jolonensis*
    CC. Staminodia erect, margin strongly involute; perianth segments rotate.—Section Stellariae.

D. Anther connective with two broad appendages on abaxial surface; occurring on serpentine. .... 8. *B. stellaris*

DD. Anther connective lacking broad appendages.

E. Perianth tube not constricted; staminodia strongly involute. ..... 9. *B. insignis*

EE. Perianth tube constricted above ovary; staminodia semi-involute.

F. Anthers obcordate; perianth tube slightly constricted; anther connective with conspicuously enlarged cells on abaxial surface; occurring on serpentine; rare. .... 10. *B. pallida*

FF. Anthers linear; perianth tube strongly constricted; anther connective without conspicuously enlarged cells on abaxial surface.

G. Scape 0.2–1.0 dm high; inner perianth segments 5–7 mm wide; style 5–6 mm long. .... 11. *B. minor*

GG. Scape 1.0–2.5 dm high; inner perianth segments 4–5 mm wide; style 7–9 mm long. .... 12. *B. purdyi*

AA. Staminodia acuminate, attenuate, or lacking; corolla salverform.—Section Filifoliae.

B. Staminodia present.

C. Staminodia subulate; perianth tube thin-membranous, splitting in fruit. 13. *B. filifolia*

CC. Staminodia cuspidate; perianth tube intact in fruit; plant monophyllous. 14. *B. kinkiensis*

BB. Staminodia lacking. .... 15. *B. orcuttii*

Species Descriptions

Section Appendiculatae Hoover, Am. Midl. Nat. 22:568. 1939

Flowers large; perianth tube cylindric, becoming brittle and splitting in fruit; perianth segments ascending recurved; staminodia elongate, involuted with undulate margins.

1. **Brodiaea appendiculata** Hoover, Madroño 4:130. 1937

Warnerville, Stanislaus Co., *Hoover 1040.* Holotype, JEPS. Isotypes, CAS, UC, US.

Corm with very coarse fibrous coat, 3.0–4.5 dm below ground level; leaves crescent-shaped in transection; scape stout, 1.0–4.5 dm tall, pedicels 4–10 cm long; lower portion of perianth tubular, 8–12 mm long, perianth throat and segments violet-purple, segments 15–25 mm long, ascending recurved; staminodia erect and linear, 8–15 mm long, edges half involuted with undulate margin, white to occasionally pale lilac, approximate to stamens; stamens 3–5 mm below tips of staminodia, filament 4–7 mm long, dilated at base, with two abaxial filiform appendages at apex, anther 3–6 mm long, channeled on abaxial side, its top 3–5 mm above stigma, tips of anther lobes clawed inward, apex of anther connective dentate; style 9–12 mm long, ovary subglobose, 5–6 mm long; in fruit perianth tube translucent and splitting. Chromosome number n = 6 (Niehaus, 1965), (fig. 6).

Distribution: (fig. 7) Sierran foothills of central California from Sutter Co. to Madera Co.; rare in Coast Ranges from Napa Co. to Santa Clara Co. Flowers April–May.

Community: Valley Grassland and Foothill Woodland, in gravelly clay soils. Below 1,800 feet.

Representative specimens: *California. Sutter Co.: Pleasant Grove, *Hoover 2215,* 2 miles S of Rio Oso, *2227.* Placer Co.: Lincoln, *4089.* Sacramento Co.: 1.25 miles NW of Scott Ranch, *Nordstrom 43;* 0.6 mile E of Bridgehouse, *Bacigalupi et al.* Calaveras Co.: 1.3 miles NNE of Buena Vista Peak, *Belshaw 1986.* San Joaquin Co.: Farmington *Hoover 1058;* road between Everett School and Bellota, *Ferris 12641.* Amador Co.: 1 mile W of Ione, *Niehaus 515.* Tuolumne Co.; Keystone, *Hoover 4069.* Stanislaus Co.: 2 miles S of Oakdale, *2027;* near La Grange on Snelling Road, *2052;* Warnerville, *1040;* Merced, *2073.* Madera Co.: E of Madera, *979;* Dalton, *3455.* Mariposa Co.: 6 miles SE of Merced Falls, *Bacigalupi 3270.* Napa Co.: mouth of Sage Canyon, *Jepson 9981.* Sonoma Co.: near Warfield Station, *M. S. Baker 3166.* San Mateo Co.: Cedro Field near Stanford University, *Bacigalupi.* Santa Clara Co.: 3.75 miles E of New Almaden, *Belshaw 2257.*

---

* The lists of representative specimens I have cited include only those which are widely distributed in herbaria for reference purposes. Others document range extensions or fill gaps in the known distribution of species, or in taxa which I interpret differently from previous treatments.

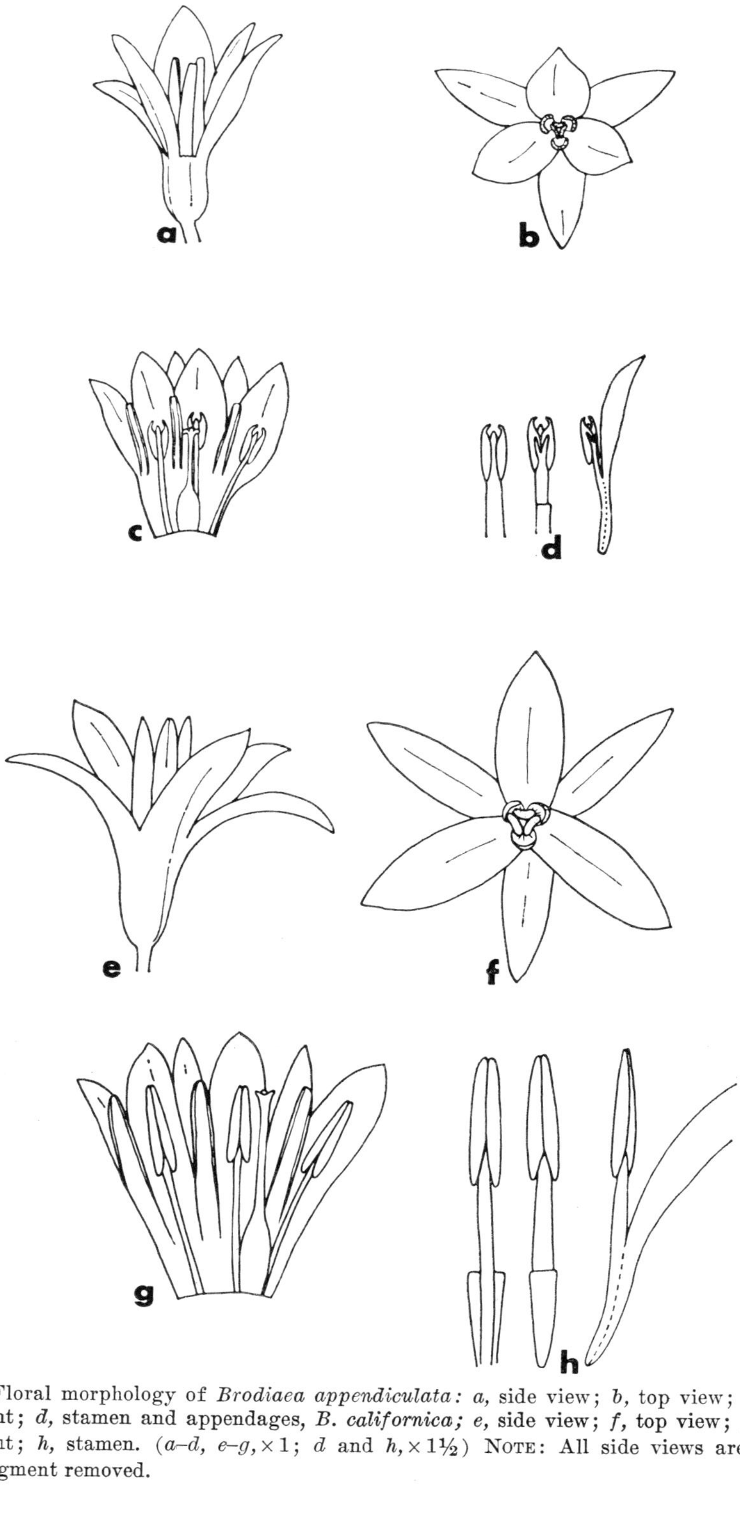

Fig. 6. Floral morphology of *Brodiaea appendiculata: a,* side view; *b,* top view; *c,* internal arrangement; *d,* stamen and appendages, *B. californica; e,* side view; *f,* top view; *g,* internal arrangement; *h,* stamen. (*a–d, e–g*, ×1; *d* and *h*, ×1½) NOTE: All side views are with one perianth segment removed.

Fig. 7. Distribution of *Brodiaea appendiculata, B. californica,* and *B. leptandra.*

Brodiaea appendiculata is found only in a narrow foothill zone from Madera County to Sutter County. It is most common in the southern portion of its distribution. It appears to grow only in stony red clay soils. The few localities cited in the Coast Ranges are the only known collections from that region. Attempts to find *B. appedicidata* in the Coast Range locations and surrounding areas failed during three different growing seasons. Since all these localities in the Coast Ranges are now under intensive agriculture, the species may now be extinct in the Coast Ranges.

2. **Brodiaea californica** Lindl., Jour. Hort. Soc. 4:84. 1849

Mountains and plains of the Sacramento, June 1848, *Hartweg 1999.*
Holotype, K. Isotypes, BM, CGE, NY.
Brodiaea grandiflora Smith var. *elatior* Baker, Jour. Linn. Soc. 11:377. 1871.
Hookera californica Greene, Bull. Calif. Acad. 2:136. 1886.

Corm with coarse fibrous coat, 1.5–3.0 dm below ground level; leaves subcrescent-shaped in transection; scape stout, 2–7 dm tall, pedicels 2–10 cm long; perianth tube 9–12 mm long, perianth throat and segments light pink to nearly white, segments 20–30 mm long, ascending recurved; staminodia erect and linear, 16–27 mm long, edges one-quarter involuted, occasionally with wavy margins, white to pale lilac, approximate to stamens; stamens equal to staminodia, filament 7–10 mm long, dilated at base, rarely with two small abaxial appendages at apex, anther 9–12 mm long, linear, slightly channeled on abaxial surface, tips of anther lobes erect with slight notch between; stigma level with top of anthers, style 18–22 mm long, ovary subglobose, 7–11 mm long; in fruit perianth tube translucent and splitting. Chromosome number n = 6 (Burbanck, 1941), (fig. 6).

Distribution: (fig. 7) Foothills of northern California from Shasta Co. to Nevada Co. Flowers late May to early July.

Community: Valley Grassland and Foothill Woodland, in gravelly clay soils. Elevation 200–2,500 feet.

Representative specimens: CALIFORNIA. Shasta Co.: 4 miles W of Ingot, *Heller 15694;* 4 miles N of Anderson, *Hoover 1212;* Redding, *Heller 7844.* Tehama Co.: 4 miles S of Corning, *Hoover 2297;* 4 miles S of Cottonwood, *1181;* SE of Red Bluff, *Crum 1451;* 14 miles E of Hwy 36 from U. S. 99, *Bacigalupi 3366.* Butte Co.: 8 miles N of Oroville, *Heller 11385;* Chico Airport *15106;* 3 miles N of Cana, *Hoover 1172.* Yuba Co.: 2 miles W of Dobbins, *Elwin 50;* 0.5 mile S of Indian Ranch, *46.* Nevada Co.: Grass Valley, *Heller 8097;* Nevada City, *Eastwood 565.*

3. **Brodiaea leptandra** (Greene) Baker, Gard. Chron. III, 20:213. 1896

Hookera leptandra Greene, Pittonia 1:74. 1887. Calistoga, Napa Co., June 1887, *C. C. Parry.* Holotype, ND-G. Isotypes, NY, US.
Hookera synandra Heller, Bull. So. Calif. Acad. 2:65. 1903. Petrified Forest–Sonoma Co., June 23, 1902, *A. A. Heller 5742.* Holotype, NY. Isotypes, DS, GH, US.

Corm 5–15 cm below ground level; leaves crescent-shaped in transection; scape stout, 2–4 dm tall, pedicels 2–5 cm long; perianth tube 6–10 mm long, perianth throat and segments rose-violet to pink, segments 18–28 mm long, outer segments 4–5 mm wide, inner 6–7 mm, ascending recurved; staminodia erect and linear, 14–16 mm long, margins half involuted and entire, white, approximate to stamens; stamens 1–2 mm below tips of staminodia, filament 8–10 mm long, slightly dilated toward base, anther 8–10 mm long, linear, slightly channeled on abaxial surface, tips of anther lobes erect with slight notch between; stigma 2–3 mm below anther apex, style 15–23 mm long, ovary subglobose, 5–7 mm long; in fruit perianth tube translucent and splitting. Chromosome number n = 6 (Niehaus, 1965), 12 (fig. 8).

Distribution: (fig. 7) California, North Coast Ranges, Lake Co., Sonoma Co., and Napa Co. Flowers in June.

Community: Mixed Evergreen Forest, in open areas. Below 3,000 feet.

Representative specimens: CALIFORNIA. Lake Co.: N side of Mt. St. Helena, *Hoover 5904.* Napa Co.: Howell Mountain, *M. S. Baker 10158.* Sonoma Co.: 2 miles SE of Healdsburg, *Hoover 3604;* Adobe Canyon near Kenwood, *M. S. Baker 2346.*

Section *Coronariae* Hoover, Am. Midl. Nat. 22:555. 1939

Flowers large; perianth tube attenuate or rounded at base, of tough texture, not splitting in fruit; perianth segments ascending recurved; staminodia strongly involute or plane.

4*a*. **Brodiaea coronaria** (Salisb.) Engler, Notizbl. Konigl. Bot. Gart. 2:317. 1899. Subspecies *coronaria*

Hookera coronaria Salisb., Parad. Lond. pl. 98. 1808. New Georgia, Northwest Coast of America (Bluffs near Fort Vancouver, Washington), 1792, *Menzies 1.* Holotype, BM. Isotype, BM.
Brodiaea grandiflora Smith, Trans. Linn. Soc. London 10:2. 1811.
Brodiaea synandra Jepson, Fl. Calif. 1:288. 1922.
Brodiaea howellii Eastw., Leafl. West. Bot. 2:111. 1938.

Corm with heavy fibrous coat, 2–15 cm below ground level; leaves crescent-shaped in transection; scape 5–25 cm tall, pedicels 1–5 cm long; perianth tube 6–13 cm long, perianth throat and segments rose-violet, segments 15–26 mm long, ascending recurved, outer segments 5–6 mm wide, inner 6–9 mm; staminodia 10–11 mm long, margins three-quarters involuted, white, approximate to stamens; stamens 2–3 mm below tips of staminodia, filament 3–4 mm long, with triangular flap on abaxial surface, anther 5–7 mm long, slightly channeled on abaxial surface, rounded notch between apical tips of anther; stigma 1–2 mm below anther apex, style 9–11 mm long, ovary ovoid, 8–9 mm long; in fruit perianth tube opaque and not splitting. Chromosome number n = 6, 12 (Niehaus, 1965), (fig. 8).

Distribution: (fig. 9) Southwest British Columbia, Canada, south to Washington, Western Oregon, northern and central California. Flowers April to June.

Community: Valley Grassland in California, prairies of western Oregon and Washington. Below 5,000 feet.

Representative specimens: BRITISH COLUMBIA. Victoria, *Connell;* Mapleguard Point between Courtney and Nanaimo, Vancouver Island, *Calder & MacKay 30083.*

WASHINGTON. San Juan Co.: Wasp Island, *Foster;* Stuart Island, *Lawrence 44;* Rosario Beach, Fidalgo Island, *Eyerdam.* Island Co.: Deception Pass State Park, Whidbey Island, *H. W. Smith 1330.* Skagit Co.: Sinclair Island, *Sprague;* near Mount Vernon, *Owen.* Kitsap Co.: Waterman, *Warren 76.* King Co.: Alki Point, *Piper 295.* Pierce Co.: Lake Park, *2093.* Thurston Co.: Gate City, *Henderson;* 8 miles E of Olympia, *Meyer 958.* Grays Harbor Co.: near McCleary, *J. W. Thompson 7350.* Yakima Co.: 6 miles NE of Bickleton, *Davis 191.* Klickitat Co.: head of Rock Creek, W of Klamath Lake, *Applegate 4417.*

OREGON. Marion Co.: Salem, *O. B. Johnson 295.* Hood River Co.: Dry Creek Falls, Cascade Locks, *Gustafson 142.* Linn Co.: Peterson Butte, *Whitaker.* Lane Co.: Pacific Hwy near border with Douglas Co., *Ingram 2077.* Jackson Co.: W of Pinehurst, *J. W. Thompson 13144.* Klamath Co.: head of Rock Creek, W of Klamath Lake, *Applegate 4417.*

CALIFORNIA. Modoc Co.: 9 miles NW of Canby, *Bacigalupi 5980.* Siskiyou Co.: Trinity Alps Camp, *M. S. Baker 2636–26.* Shasta Co.: *Stevens, 358;* Cottonwood, *Hoover 1185;* Sisson (City of Mt. Shasta), *Setchell & Dobie;* 2 miles W of Fall River Mills, *Hitchcock 6625.* Lassen Co.: 5 miles S of Adin, *Cronquist 8485;* Dixey Mountain, *Baker & Nutting;* Black Mountain, *Hormay.* Butte Co.: near Clear Creek School, *Heller 11382;* 10 miles NW of Chico, *Hoover 2246.* Sutter Co.: Pleasant Grove, *1145;* Riego, *1141.* Sacramento Co.: Sacramento, *1132;* Fair Oaks, *Ramaley 11257;* 7.5 miles W of Ione, *Niehaus 512.* Calaveras Co.: Salt Springs Valley, *Tracy 5662;* near Milton, *Davy 1346.*

4b. **Brodiaea coronaria** subsp. **rosea** (Greene) Niehaus, comb. nov.

Hookera rosea Greene, Bull. Calif. Acad. 2:137. 1886. Near Houghs Springs (Lake Co.), May 1884, *M. K. Curran.* Holotype, US.
Brodiaea rosea Baker, Grad. Chron. III. 20:214. 1896.

Description generally as for subspecies *coronaria,* but with the following differences: Corm with thin fibrous coat, 5–7 cm below ground level; scape 4–7 cm tall, pedicels 1–4 cm long; perianth tube 7–9 mm long, perianth throat and segments rose to light pink, segments 12–18 mm long, outer segments 3–5 mm wide, inner 4–6 mm; staminodia white to pink; stamen filament 2–3 mm long; style 6–8 mm long, ovary ovoid, 6–8 mm long. Chromosome number n = 6 (Niehaus, 1965), (fig. 10).

Distribution: (fig. 9) CALIFORNIA, known from only three stations in inner North Coast Ranges of Tehama Co. and Lake Co. Flowers late May to June.

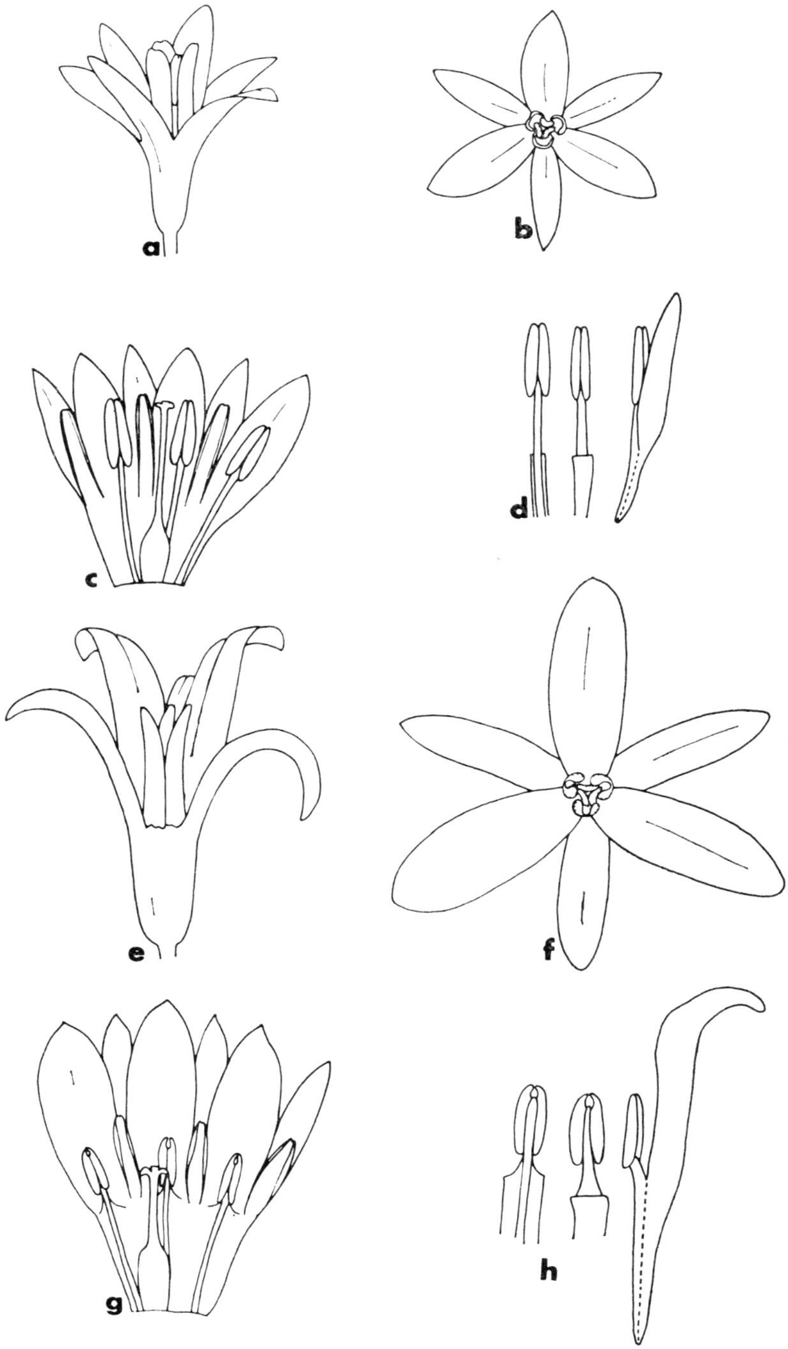

Fig. 8. Floral morphology of *Brodiaea leptandra: a,* side view; *b,* top view; *c,* internal arrangement; *d,* stamen. *B. coronaria* subsp. *coronaria: e,* side view; *f,* top view; *g,* internal arrangement; *h,* stamen. (*a–c,* × 1; *d–g,* × 1½; *h,* × 2)

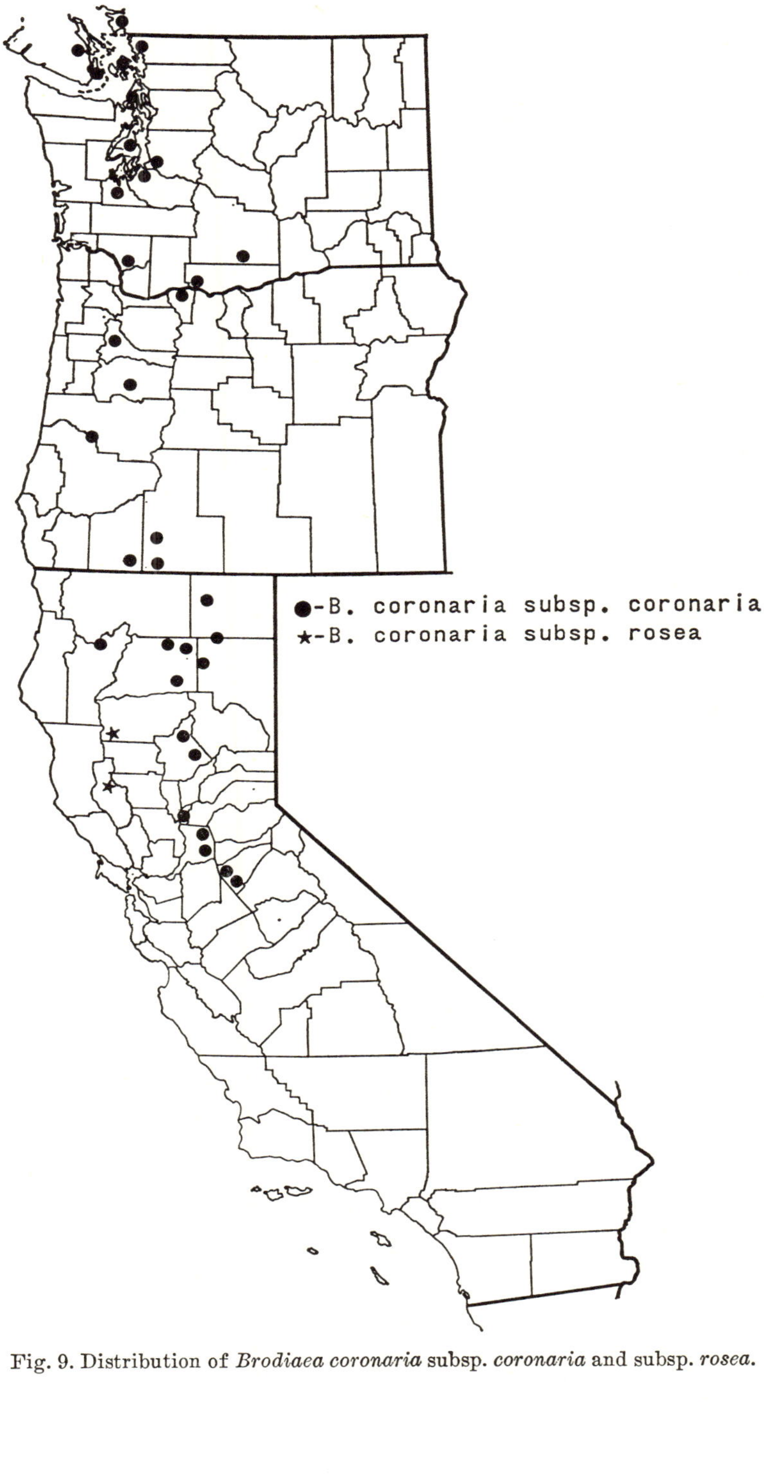

Fig. 9. Distribution of *Brodiaea coronaria* subsp. *coronaria* and subsp. *rosea*.

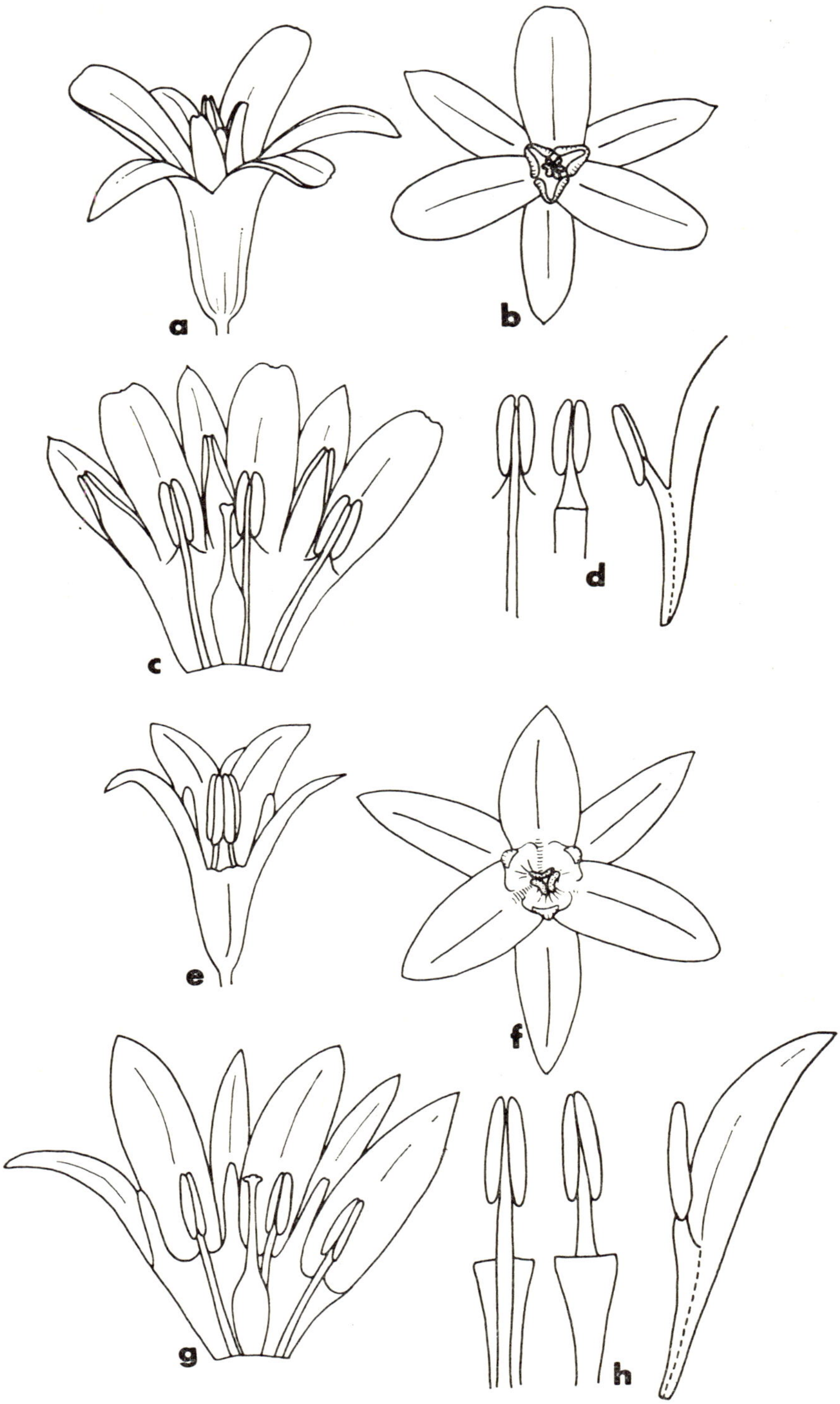

Fig. 10. Floral morphology of *Brodiaea coronaria* subsp. *rosea: a,* side view; *b,* top view; *c,* internal arrangement; *d,* stamen. *B. elegans* subsp. *elegans: e,* side view; *f,* top view; *g,* internal arrangement; *h,* stamen. (*a–c* and *e–g,* × 1½; *d* and *h,* × 2)

Community: Oak Woodland, in open areas on serpentine soil. Elevation 1,000–5,000 feet.

Representative specimens: CALIFORNIA. Tehama Co.: serpentine 10.7 miles W of Paskenta, *Hoffman 3692;* logging road W of Paskenta at Whiskey Saddle, 4,753 feet, *Baker & Wagnon 12797.* Lake Co.: head of Indian Valley near stream crossing, *Niehaus 556.*

Brodiaea coronaria subsp. *rosea* grows only on serpentine soil. The population at the head of Indian Valley, Lake County will be inundated from a new reservoir that will fill the valley. A new population has been found at the head of Stanton Creek which will be immediately above the reservoir waterline. The flowers of collections from Tehama County vary somewhat from those of Indian Valley.

5a. **Brodiaea elegans** Hoover, Am. Midl. Nat. 22:555. 1939. Subspecies *elegans*

3 miles N of Nevada City, Nevada Co., Calif., *Constance 2289.* Holotype, UC. Isotypes, GH, RSA, US, WTU.

Brodiaea coronaria Jepson var. *mundula* Jepson, Fl. Calif. 1:287. 1922.

Brodiaea elegans var. *mundula* Hoover, Am. Midl. Nat. 22:557. 1939.

Corm 5–15 cm below ground level; leaves crescent-shaped in transection; scape stout, 1–5 dm tall, pedicels 0.5–10.0 cm long; perianth tube funnelform, 8–19 mm long, perianth throat and segments purple-violet, segments 15–30 mm long, ascending recurved, outer segments 4–7 mm wide, inner 6–10 mm; staminodia erect and flat, 6–9 mm long, tip plain to trilobed, white to pale lilac, distant from stamens; stamens equal or slightly below staminodia, filament 4–6 mm long, not dilated at base, with narrow linear wings on abaxial surface, anther 4–10 mm long, slightly channeled on abaxial surface, rounded notch between apical tips of anther; stigma 3–6 mm below anther apex, style 7–15 mm long, ovary ovoid, 9–15 mm long; in fruit perianth tube opaque and not splitting. Chromosome number n = 6, 8 (Niehaus 1965), 12, 16 (Burbanck, 1941), (fig. 10).

Distribution: (fig. 11) Western Oregon, Coos Co. and Josephine Co.; northern and central California, in Coast Ranges to San Benito Co. and in Sierra Nevada foothills to Kern Co. Flowers May to July.

Community: Valley Grassland, Foothill Woodland, and Yellow Pine Forest. From sea level to 7,000 feet.

Representative specimens: OREGON. Coos Co.: E slope of Iron Mountain, *W. H. Baker 3368.* Josephine Co.: Savage Creek near Grants Pass, *Henderson 7202.*

CALIFORNIA. Del Norte Co.: 3 miles E of Gasquet, *Tracy 12723.* Humboldt Co.: Grouse Mountain, *13014;* Carlotta, *17615;* South Yager Creek, *6780;* Larrabee Valley, *8770;* Garberville, *17163;* Alderpoint, *18867;* Eureka, *16060.* Siskiyou Co.: Oro Fino, *Butler 891;* Yreka, *1421;* White Gulch, Salmon Mountains, *Tebbe 135.* Trinity Co.: Stuart Fork of Trinity River, *Kellogg 5544;* Big Bar, *Clar 88;* Ruth, *Tracy 8800.* Mendocino Co.: Ukiah, *Purdy;* summit of Covelo Hwy, *M. S. Baker 10372.* Lake Co.: Cobb Mountain, *Tracy 14015;* Butts Canyon, *Sweeny 961;* Elk Mountain, *Tracy 19609;* Indian Valley, *Hoover 3577;* Kelsey, *Jones.* Colusa Co.: Stoney Creek, *Sharsmith 4268.* Sonoma Co.: Healdsburg, *Hoover 3606;* Sonoma Creek Canyon, *M. S. Baker;* Sonoma, *Rose 40363.* Napa Co.: Napa, *Sharsmith 5110.* Marin Co.: Tomales Bay, *Schreiber 514;* Mt. Tamalpais, *Setchell;* San Anselmo Canyon, *Howell.* Solano Co.: Gates Canyon, *Lewis 92.* Alameda Co.: Berkeley, *Michener & Bioletti 2066a.* Contra Costa Co.: Mt. Diablo, *Bowerman 229;* Niles, *Jepson.* Santa Clara Co.: San Antonio Valley, *Sharsmith 3292;* Los Gatos, *Heller 7442;* New Almaden, *Belshaw 2260.* Santa Cruz Co.: Slate Creek, *Carlson 72;* Capitola, *Belshaw 2297.* San Benito Co.: 2.25 miles N of San Juan Bautista, *2190;* 1.25 miles N of Sugarloaf Peak, *2177.* Shasta Co.: Montgomery Creek, *Eastwood 663;* Hooten Gulch (Redding), *Johansen 95.* Butte Co.: 8 miles N of Oroville, *Heller 11385;* near Paradise, *16105; Chico Airport, 15105.* Nevada Co.: Kress, *Hall 10205;* Grass Valley, *Niehaus 128.* Placer Co.: Auburn, *Lemmon;* Newcastle, *Clar 127;* Colfax, *Nordstrom 144;* Sheridan, *Hoover 4094.* El Dorado Co.: Plum Creek, *Johansen 345;* 1.5 miles NE of Placerville, *Belshaw 2335.* Sacramento Co.: Sacramento, *Copeland 1569.* Amador Co.: Ione, *Braunton 1009;* Sutter Creek, *Nordstrom 797.* Calaveras Co.: Copperopolis, *Davy 51013;* Salt Springs Valley, *Hoover*

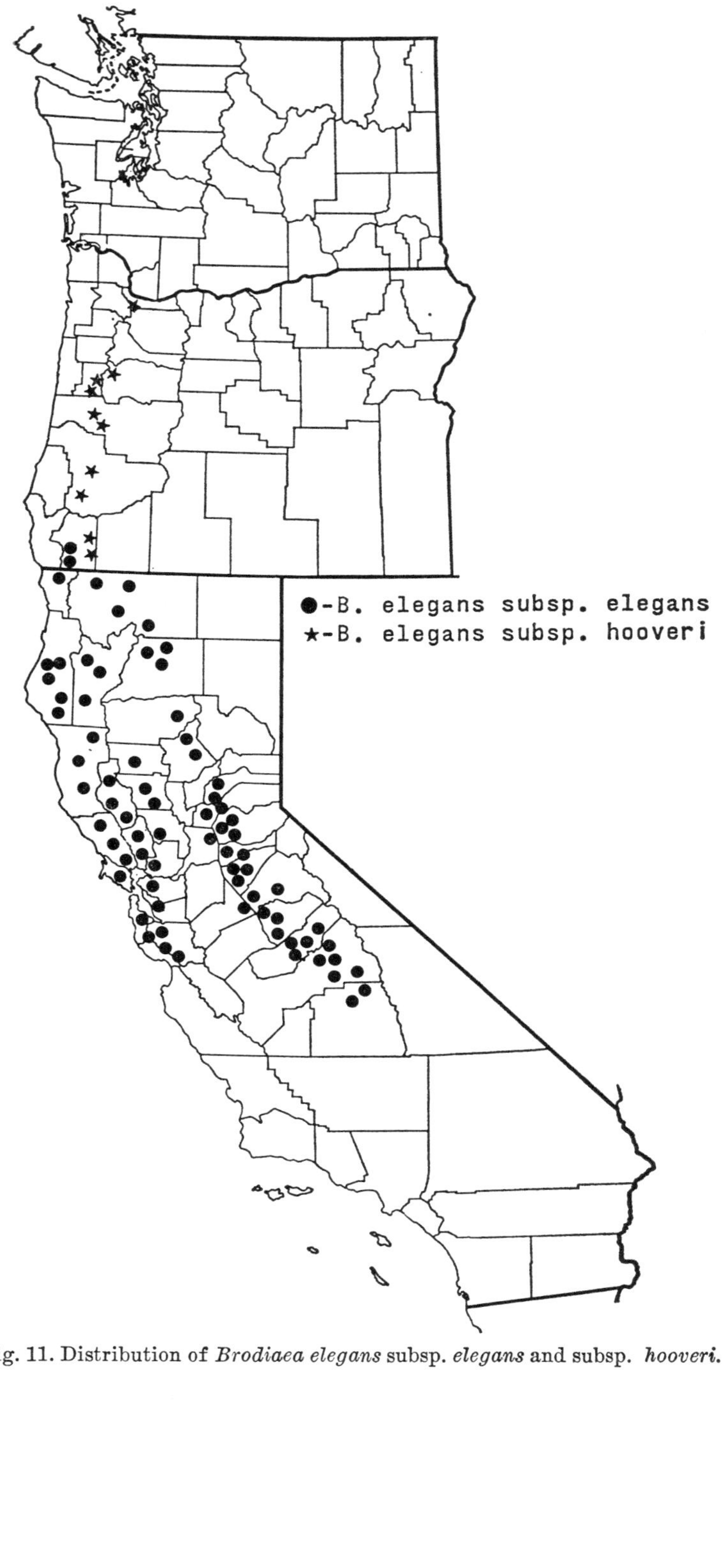

Fig. 11. Distribution of *Brodiaea elegans* subsp. *elegans* and subsp. *hooveri*.

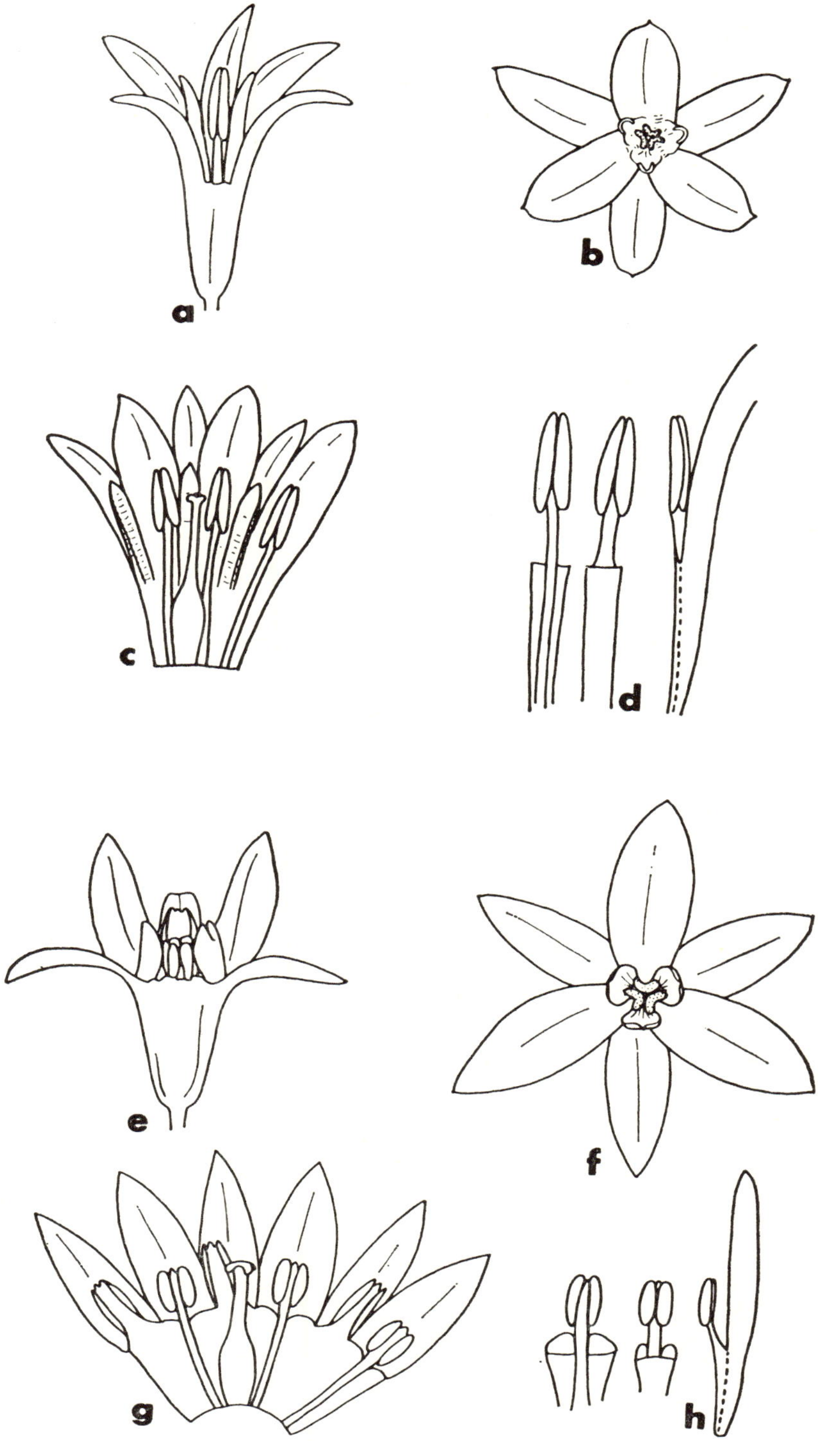

Fig. 12. Floral morphology of *Brodiaea elegans* subsp. *hooveri: a,* side view; *b,* top view; *c,* internal arrangement; *d,* stamen. *B. terrestris* subsp. *terrestris: e,* side view; *f,* top view; *g,* internal arrangement; *h,* stamen. (*a–c* and *e–g,* × 1½; *d* and *h,* × 2)

2438; Milton, *Davy 1323.* Tuolumne Co.: N of Columbia, *Wiggins 14358;* Hog Ranch near Mather, *Straw 768;* Chinese Camp, *Rutter 242;* Groveland, *Hoover 4129,* 3 miles E of Sonora, *3618.* Stanislaus Co.: Red Mountain, *Sharsmith 3110;* Adobe Valley, *Elmer 4333;* Warnerville, *Hoover 2045;* 10 miles S of Modesto, *573.* Merced Co.: 10 miles E of Merced, *1118.* Mariposa Co.: Coulterville, *Wolf 5123.* Madera Co.: 1 mile N of Friant, *Bacigalupi 10372;* 1.5 miles S of Miami Mountain, *Schlobohm 11239.* Fresno Co.: Orange Cove, *Hoover 1008;* Pinedale, *989;* Auberry, *Bullard 17;* Mist Falls, S fork of Kings River, *Niehaus 828.* Tulare Co.: trail to Pinto Lake from Cliff Creek, *Lorraine 10932;* Three Rivers, *Niehaus 677.*

5b. **Brodiaea elegans** subsp. **hooveri** Niehaus, subspecies nov.

1 mile E on McKenzie highway from junction with Coburg road on bluff above McKenzie River, Lane Co., Oregon, *Niehaus 696.* Holotype, UC. Isotype, US.

Perianthii tubus 13–17 mm longus, segmenta eius 22–30 mm longa, ea exteriora 4–5 mm, ea interiora 6–9 mm lata; staminodia erecta, apice reflexa, marginibus quartam partem latitudinis eorum involutis; antherae 6–8 mm longae; stylus 12–15 mm longus, ovarium ovoideum, 9–11 mm longum; chromosomatum numerus = 20.

Description generally as for subspecies *elegans,* but with the following differences: Perianth tube 13–17 mm long, segments 22–30 mm long, outer segments 4–5 mm wide, inner 6–9 mm; staminodia erect with tip reflexed, margin one quarter involuted; anther 6–8 mm long; style 12–15 mm long, ovary ovoid, 9–11 long. Chromosome number n = 20 (fig. 12).

Distribution: (fig. 11) Western Oregon, Clackamas Co. to Lane Co. and Jackson Co.

Representative specimens: OREGON. Clackamas Co.: west bluffs, Willamette River Falls, Oregon City, *Constance & Beetle 2773.* Linn Co.: dry hills beyond Holly, *Henderson 13606.* Benton Co.: W of Corvallis, *Gilbert 481.* Lincoln Co.: Cape Perpetua, *Detling 5011.* Lane Co.: Florence road, *Henderson 16786;* Crow road near Eugene, *Detling 4925.* Jackson Co.: Little Butte Creek, *8069;* Sams Valley, *Henderson 12779.*

Brodiaea elegans subsp. *hooveri* is found in the Willamette Valley of Oregon and southward near Ashland. The staminodia vary from slightly involute to strongly so in all the populations that I have observed.

Section Terrestres Niehaus, section nov.

Flores parvi; perianthii tubus campanulatus, textura tenace, apud fructum haud findens, segmenta eius adsendentia recurvataque; staminodia brevia, leviter involuta, cucullo partiali intus laeve inclinato apici affini instructa.

Flowers small; perianth tube campanulate, of tough texture, not splitting in fruit, perianth segments ascending recurved; staminodia short, slightly involute with a partial hood along apex, leaning inward slightly.

6a. **Brodiaea terrestris** Kellogg, Proc. Calif. Acad. 2:6. 1895.
Subspecies *terrestris*

Brodiaea grandiflora Smith var. *macropoda* Torr., Pacif. Railroad Rep. 4:149. 1857. Swamps, Santa Rosa Creek and Laguna, Calif., *Bigelow.* Holotype, K. Isotype, NY.

Brodiaea torreyi Wood, Proc. Acad. Phila. 20:172. 1868.

Hookera terrestris Britten, Jour. Bot. 24:51. 1886.

Hookera macropoda Ktze., Rev. Gen. 2:712. 1891.

Brodiaea coronaria var. *macropoda* (Torr.) Hoover, Am. Midl. Nat. 22:562. 1939.

Corm 2–8 cm below ground level; leaves cresent-shaped in transection; scape 0.5–7.0 cm tall, pedicels 3–15 cm long; perianth tube 6–9 mm long, perianth throat and segments rose-violet, segments 10–16 mm long, ascending recurved, outer segments 4–6 mm wide, inner 5–7 mm; staminodia leaning inward slightly, margin one-quarter involuted, becoming slightly hooded at apex, violet to nearly white, distant from anthers; stamens 2–3 mm below apex of staminodia, filament 2–3 mm long, not dilated, with narrow triangular flap on abaxial surface, anther 3–4 mm long, channeled on abaxial surface, apex of anther connective dentate; stigma

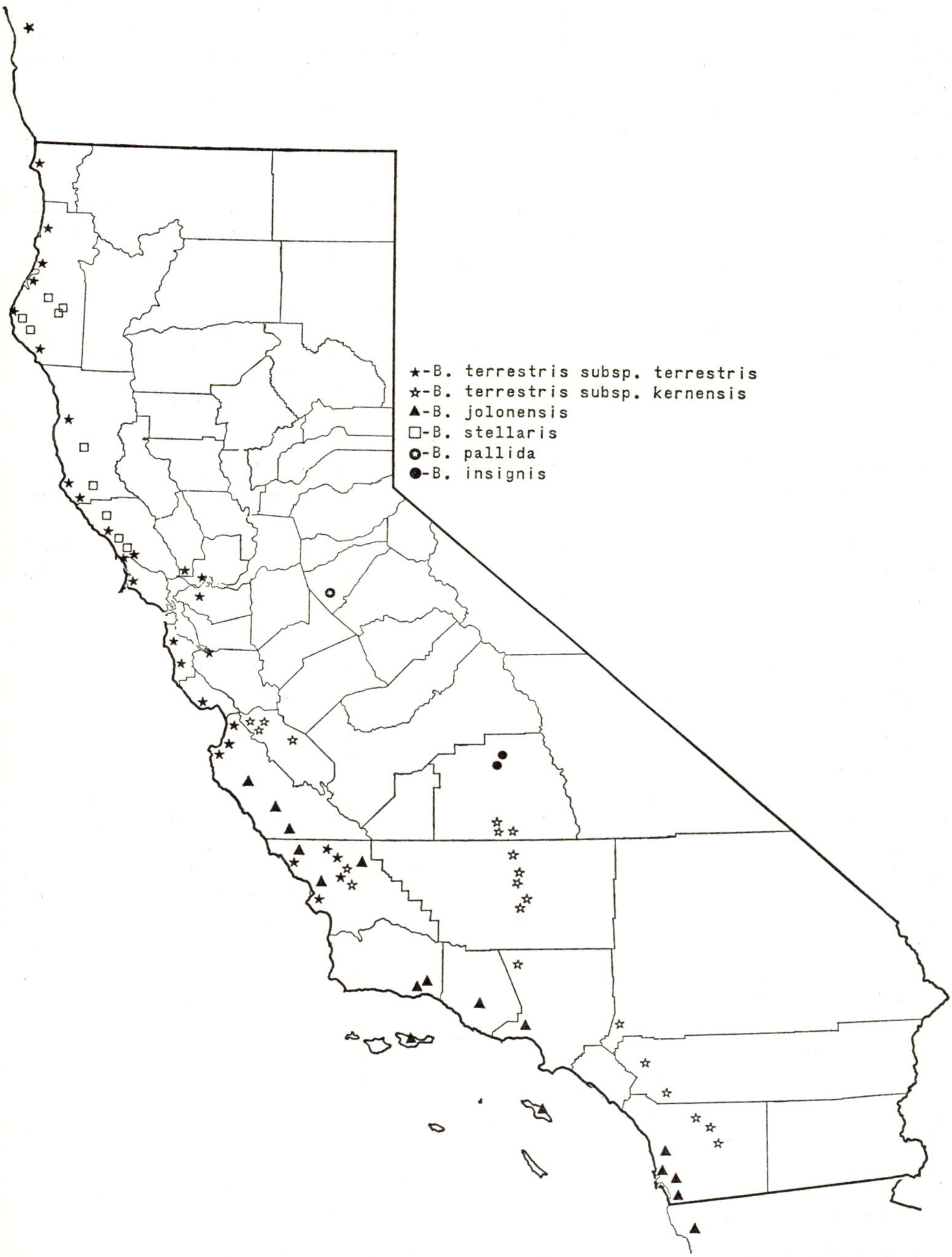

Fig. 13. Distribution of *Brodiaea terrestris, B. jolonensis, B. stellaris, B. pallida,* and *B. insignis.*

slightly below anther apex, style 4–5 mm long, ovary ovoid, 7–9 mm long; in fruit perianth tube opaque and not splitting. Chromosome number n = 6, 18 (Niehaus, 1965), (fig. 12).

Distribution: (fig. 13) Pacific Coast from Coos Bay, Oregon, to San Simeon, California. Flowers April to June.

Community: Coastal Prairie, Foothill Woodland. From sea level to 1,500 feet.

Representative specimens: OREGON. Coos Co.: Charleston, *Spaffer.* CALIFORNIA. Del Norte Co.: Crescent City, *Parks 8274.* Humboldt Co.: Trinidad, *Tracy 8111;* Mouth of Little River, *Wiggins 5877;* Eureka, *Tracy 3244;* Cape Mendocino, *14893.* Mendocino Co.: Fort Bragg, *Bolander 4759;* Havens Neck, *Niehaus 674;* Gualala, *M. S. Baker 8995.* Sonoma Co.: Pitkin Marsh, *3002j;* Bodega, *Chandler 672.* Marin Co.: Point Reyes, *Davy 6848;* Drakes Beach, *Niehaus 535;* 9 miles W of Novato, *Howell 16296.* Napa Co.: Napa Junction, *Sonne.* Solano Co.: S of Elmira, *Heller 15598.* Contra Costa Co.: Byron Hot Springs, *Hoover 2122;* 1.5 miles E of Mulligan Hill, *Belshaw 16027.* San Mateo Co.: Crystal Springs, *Eastwood 334;* Burlingame, *Rose 35159;* Moss Beach, *Brandegee.* Santa Clara Co.: Milpitas, *Smith.* San Benito Co.: between San Benito River and Bitterwater, *Hall 9908.* Santa Cruz Co.: 1.5 miles N of Terrace Point, *Wilson 181.* Monterey Co.: SW of Sugarloaf Peak, *Belshaw 2179A;* Pacific Grove, *Eastwood 104;* Point Lobos, *Mason 9355;* San Antonio River, *Brennen 560.* San Luis Obispo Co.: Paso Robles, *Condit;* near Santa Margarita, *Hoover 4080;* NE of Chimney Rock, *Nordstrom 1125;* Point Sierra Nevada, *Hoover 9100;* Pismo, *Unangst.*

6b. **Brodiaea terrestris** subsp. **kernensis** (Hoover) Niehaus, comb. nov.

Brodiaea coronaria Engler var. *kernensis* Hoover, Am. Midl. Nat. 22:561. 1939. Poso Creek Oil Field, Kern Co., Calif., *Hoover 4083.* Holotype, UC. Isotypes, DS, NY, US.

Brodiaea elegans var. *australis* Hoover, Leafl. West. Bot. 8:132. 1957. Terra Bella, Tulare Co., *Hoover 8372.* Holotype, OBI.

Description generally as for subspecies *terrestris,* but with the following differences: Leaves subcrescent-shaped; scape 2–20 cm tall, pedicels 1–7 cm long; perianth tube 11–13 mm long, segments 15–20 cm long, outer segments 5–6 mm wide, inner 6–9 mm; anther 5–6 mm long; style 7–9 mm long, ovary ovoid, 8–9 mm long. Chromosome number n = 24 (fig. 14).

Distribution: (fig. 13) South Coast Ranges to Transverse and Peninsular Ranges of southern California, Sierra Nevada foothills from Tulare Co. southward. Flowers April to June.

Community: Valley Grassland, Foothill Woodland. From near sea level to 5,000 feet.

Representative specimens: CALIFORNIA. San Benito Co.: E of Panoche Pass, *Hoffman 1405;* Hollister, *Hoover 3486;* 2.5 miles NE of San Juan Bautista, *Belshaw 2184.* Monterey Co.: 2 miles SW of Aromas, *2152.* San Luis Obispo Co.: 17 miles E of Creston, *Hoover 7865;* 2.75 miles NW of El Dorado School, *Hendrix 246.* Ventura Co.: Sivernale Ranch near Chatsworth, Santa Susanna Mountains, *Kappler;* near Ojai, *Hood 41-88K.* Tulare Co.: Terra Bella, *Hoover 8372;* 0.85 mile E of Fountain Springs, *Niehaus 746;* 4.35 miles E of Ducor, *747.* Kern Co.: opposite Garces historical monument, Garces-Woody road, *631;* Oil City, *Heller 7637;* Tejon Creek, *Keck & Clausen 3168;* 1.5 miles NE of Kernville, *Voegelin 195.* San Bernadino Co.: Squints Ranch, *Anderson 6537.* Los Angeles Co.: Elizabeth Lake, *Parish 1931.* Riverside Co.: Strawberry Valley, *Hall 2093;* near Murrieta, *Lathrop.* San Diego Co.: Laguna Lakes region, Laguna Mountains, *Pierson 4834;* Palomar Mountain, *4811;* SW side of Cuyamaca Lake, *Wiggins 3173.*

Brodiaea terrestris subspecies *kernensis* is common and widespread. Previously all specimens from localities outside Tulare and Kern counties have been identified as *B. coronaria.* This subspecies appears to be an octoploid version of *B. terrestris.* Populations examined in Tulare County vary in stature, with two populations having tall scapes typical of subspecies *kernensis.* One population east of Fountain Springs has short scapes, which are typical of *B. terrestris* subsp. *terrestris,* but with the larger flowers of subsp. *kernensis.*

7. **Brodiaea jolonensis** Eastw., Leafl. West. Bot. 2:111. 1938

Near Jolon, 1915, *Starr.* Holotype, CAS.

Corm 2–15 cm below ground level; leaves crescent-shaped in transection; scape 0.5–1.5 dm tall, pedicels 1–4 cm long; perianth tube 7–9 mm long, perianth throat and segments violet, segments 11–18 mm long, ascending recurved, outer segments 4–6 mm wide, inner 6–8 mm;

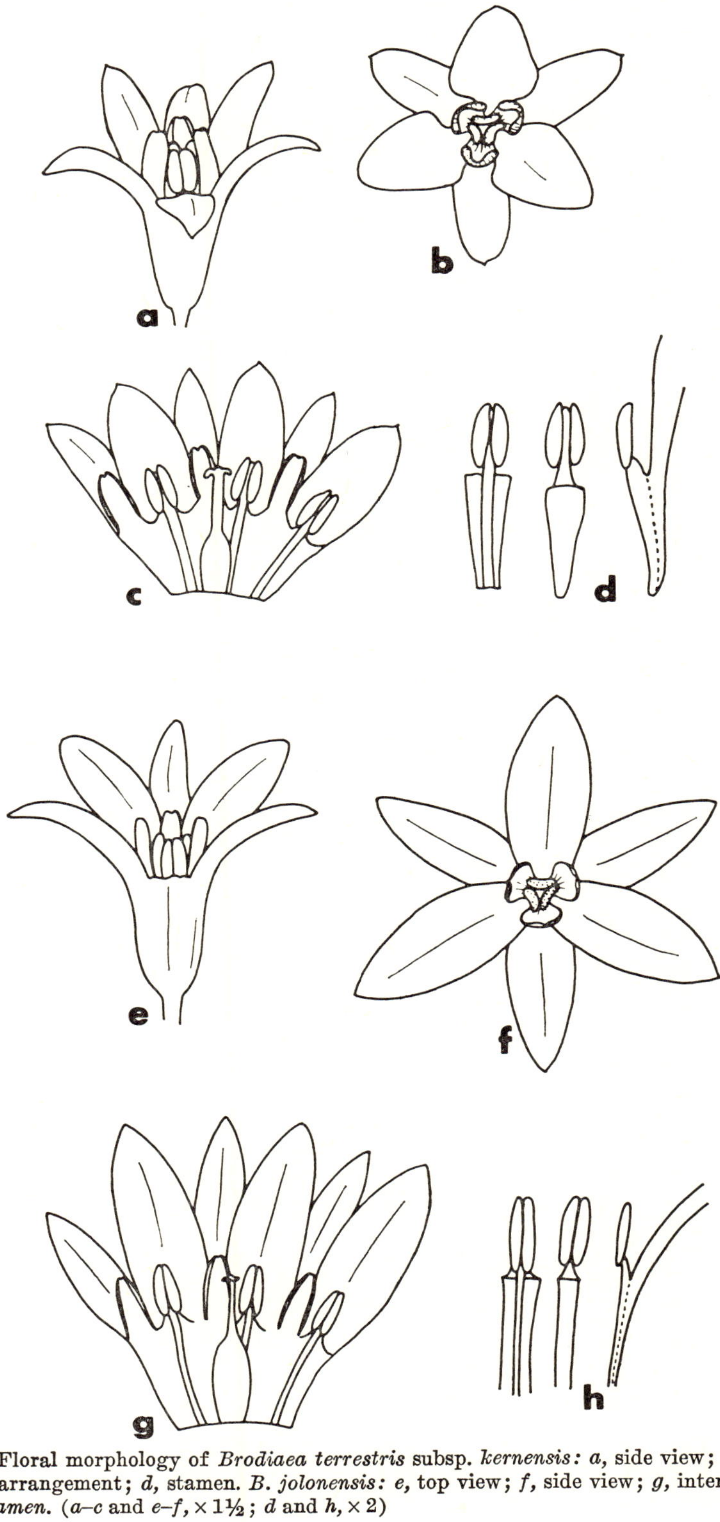

Fig. 14. Floral morphology of *Brodiaea terrestris* subsp. *kernensis: a,* side view; *b,* top view; *c,* internal arrangement; *d,* stamen. *B. jolonensis: e,* top view; *f,* side view; *g,* internal arrangement; *h, stamen.* (*a–c* and *e–f*, × 1½; *d* and *h*, × 2)

staminodia leaning inward slightly, 5–6 mm long, margin one-quarter involuted, becoming slightly hooded at apex, violet, with apex near stamens; stamens slightly below staminodia, filament 1–2 mm long, with narrow bilobed triangular wing on abaxial surface, violet; anther 4–5 mm long, with V-shaped notch at apex, tips clawed inward; stigma slightly below anther apex, style 5–7 mm long, ovary ovoid, 5–6 mm long, light purple; in fruit perianth tube thick and opaque, not splitting. Chromosome number n = 6 (Niehaus 1965), 18 (fig. 14).

Distribution: (fig. 13) California, from coastal Monterey Co. to San Diego Co. and adjacent Baja California, Mexico. Flowers April to May.

Community: Valley Grassland, Foothill Woodland, Coastal Prairie, on clay flats. From sea level to 1,000 feet.

Representative specimens.—CALIFORNIA. Monterey Co.: 0.5 mile N of Jolon, *Niehaus 511;* Tassajara Hot Springs, *Elmer 3218.* San Luis Obispo Co.: Morro Beach, *Condit.* Santa Barbara Co.: between Junction Camp and Ranger Peak, *Chandler 3106;* Del Playa Drive, Isla Vista, *Niehaus 509;* El Crielete Road, E of Santa Barbara, *508;* NW of ranch, Santa Rosa Island, *Hoffman 722;* 0.25 mile W of Portezuela Grade, Santa Cruz Island, *C. F. Smith 6121.* Los Angeles Co.: Topanga Canyon, *Taylor 3382;* W of Cornell Corners, Santa Monica Mountains, *Raven & Verity 15350;* 3.5 miles from Avalon on road to Isthmus, Santa Catalina Island, *Wolf 3623.* San Diego Co.: Rancho Santa Fe, *Gander 8309;* Kearney Mesa, *8317;* Otay Mesa, *Niehaus 662.* MEXICO. BAJA CALIFORNIA. NE of Saujal, *Wiggins 10089.*

Section Stellares Hoover, Am. Midl. Nat. 22:564. 1939.

Flowers small; perianth tube rounded or somewhat truncate at base, of tough texture, persistent and not splitting in fruit; perianth segments rotately spreading; staminodia broad, involuted, obtuse to emarginate.

8. **Brodiaea stellaris** S. Wats., Proc. Am. Acad. 17:381. 1882

Plentiful on high mountain in vicinity of Ukiah, California, June 1881, *Carl Purdy.* Holotype, GH. Fragment of type, JEPS.

Corm 5–10 cm below ground level; leaves crescent-shaped in transection; scape 0.2–0.6 dm tall, pedicels 1–5 cm long; perianth tube campanulate, 7–10 mm long, perianth throat and segments violet, segments 7–15 mm long, ascending recurved, outer segments 4–5 mm wide, inner 4–7 mm; staminodia erect and broad, with margin one-quarter involuted, apex broad, white, approximate to stamens; stamens 1–2 mm below apex of staminodia, filament 1–3 mm long, with two broad wings on abaxial surface of anther connective, anther 4–6 mm long, deeply channeled on abaxial surface, anther connective dentate at apex, apical tips of anther lobes clawed inward; stigma 2–3 mm below anther apex, style 4–5 mm long, ovary ovoid, 6–9 mm long; in fruit perianth tube transparent and not splitting. Chromosome number n = 6 (Johansen, 1932), (fig. 15).

Distribution: (fig. 13) California, North Coast Ranges from Humboldt Co. to Sonoma Co. Flowers May–June.

Community: Mixed Evergreen Forest, Redwood Forest, in open areas on serpentine. From sea level to 3,000 feet.

Representative specimens.—CALIFORNIA. Humboldt Co.: McClellan Mountain, *Tracy 18903;* Fort Baker Ranch, *10270;* 1 mile W of Petrolia, *18766;* mouth of Davis Creek, S of Cape Mendocino, *18756.* Mendocino Co.: near Ukiah, *Eastwood 11489.* Sonoma Co.: McMurray Ranch, Healdsburg-Guerneville Hwy, *Hoffman 3595;* Harrisburg Grade, *Hoover 3611.*

9. **Brodiaea insignis** (Jepson) Niehaus, comb. nov.

Brodiaea synandra var. *insignis* Jepson, Fl. Calif. 1:288. 1922. Shepherd Cove (Tulare Co.), May 25, 1915, *Fry 397.* Holotype, JEPS.

Corm 5–15 cm below ground level; leaves crescent-shaped in transection; scape 0.7–2.5 dm tall, pedicels 2–9 cm long; perianth tube 6–9 mm long, perianth throat and segments rose to pink, segments 11–15 cm long, rotate, outer segments 4–6 mm wide, inner 4–8 mm; staminodia erect, 6–8 mm long, margin three-quarters involuted, tip bilobed, white, approximate to stamens;

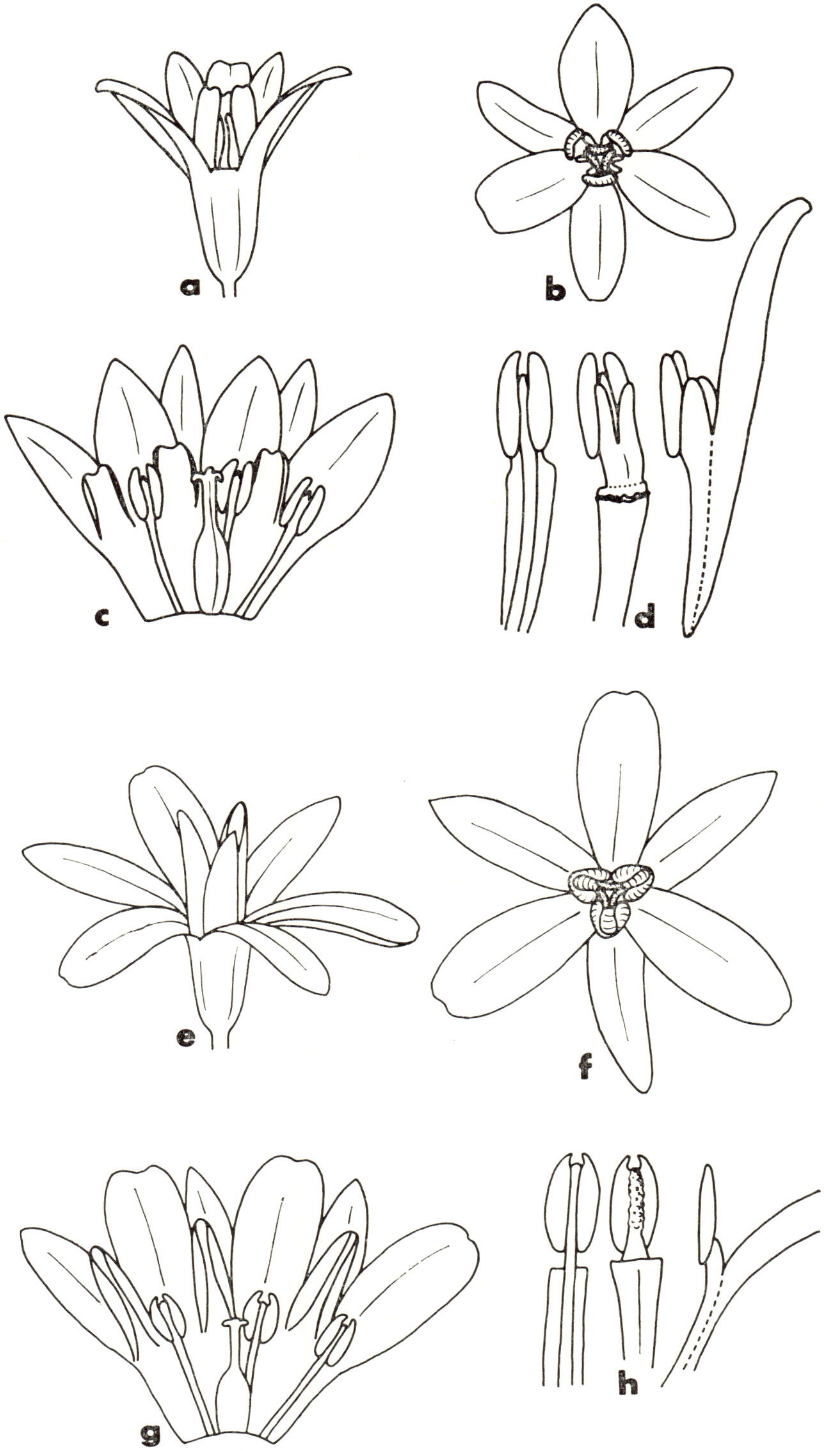

Fig. 15. Floral morphology of *Brodiaea stellaris: a,* side view; *b,* top view; *c,* internal arrangement; *d,* stamen and appendages. *B. insignis: e,* side view; *f,* top view; *g,* internal arrangement; *h,* stamen. (*a–c* and *e–g,* × 1½; *d* and *h,* × 2)

stamens 2–4 mm below apex of staminodia, filament 1–2 mm long, not dilated at base, triangular wing on abaxial surface, anther connective dentate at apex, apical tips of anther clawed inward; stigma 1–3 mm below anther apex, style 4–5 mm long, ovary ovoid, 4–5 mm long; in fruit perianth tube opaque and not splitting. Chromosome number n = 16 (fig. 15).

Distribution: (fig. 13) California, known from a few stations in Kaweah River and Tule River drainage in Tulare Co., possibly in Kern Co. also. Flowers mid May–June.

Community: Foothill Woodland, in open areas. Elevation 800–1,500 feet.

Representative specimens: CALIFORNIA. Tulare Co.: Shepherd Cove, *Fry 397;* 0.1 mile N of Salt Creek fire road on South Fork of Kaweah road, *Niehaus, Bacigalupi, & Pusateri 753;* 5.5 miles S from site of old (town) Three Rivers on S fork of Kaweah River, *754:* 4.7 miles S from old (town) Three Rivers, *755;* S fork of Tule River, *Munz 12283.*

Brodiaea insignis was long known from a single collection. For many years unsuccessful attempts were made to relocate this species. Although I was unable to find *B. insignis* at the type locality, Shepherd Cove, I did find large populations in a nearby canyon of the Kaweah River. Subsequently a herbarium specimen was found representing a population on the South Fork of the Tule River.

10. **Brodiaea pallida** Hoover, Leafl. West. Bot. 2:129. 1938

Chinese Camp, *Hoover 2375.* Holotype, UC. Isotypes, CAS, DS, NY, US.

Corm 2–5 cm below ground level; leaves crescent-shaped in transection; scape 1–2 dm tall, pedicels 0.5–3.0 cm long; perianth tube 9–11 mm long, slightly constricted at throat, perianth throat and segments rose-pink, segments 9–11 mm long, ascending recurved to nearly rotate, outer segments 3–5 mm wide, inner 4–6 mm long; staminodia erect, 8–11 mm long, middle portion of margin half involuted, tip deeply bilobed, white, approximate to stamens; stamens 3–5 mm below apex of staminodia, filament 4–5 mm long, with narrowly linear wing on abaxial surface, anther 2–3 mm long, obcordate, broad V-shaped notch at apex with apical tips of anther lobes slightly clawed inward; stigma 1–2 mm below anther apex, style 8–11 mm long, ovary subglobose, 4–5 mm long; in fruit perianth tube thin, opaque, not splitting. Chromosome number n = 6 (Niehaus 1965), (fig. 16).

Distribution: (fig. 13) California, known only from a single station 0.5 mile W of Chinese Camp, Tuolumne Co. Flowers late May to early June.

Community: Foothill Woodland, in open area along intermittent streambed. Serpentine soil. Elevation 1,260 feet.

Representative specimens: CALIFORNIA. Tuolumne Co.: W of Chinese Camp, *Hoover 2375, Niehaus 526.*

Brodiaea pallida forms a single population 10–20 feet wide for approximately one-fourth of a mile along the stream. Various hybrids can be found here between *B. pallida* and *B. elegans,* a species common on drier ground. If *B. pallida* occurs elsewhere it will probably be where there is the combination of serpentine soil and an intermittent steam.

11. **Brodiaea minor** (Benth.) S. Wats., Proc. Am. Acad. 14:236. 1879

Brodiaea grandiflora Smith var. *minor* Benth., Pl. Hartw. 340, 1857. Pine Creek, E Tehama Co., *Hartweg 2002/300.* Holotype, K. Isotypes, BM, CGE.

Hookera minor Britten, Jour. Bot. 24:51. 1886.

Brodiaea nana Hoover, Leafl. West. Bot. 1:255. 1936. 5 miles N of Snelling, Merced Co., *Hoover 963.* Holotype, CAS. Isotype, UC.

Brodiaea minor var. *nana* Hoover, Am. Midl. Nat. 22:566. 1939.

Corm 2–8 cm below ground level; leaves crescent-shaped in transection; scape 0.2–1.0 dm tall, pedicels 1–3 cm long; perianth tube 6–10 mm long, strongly constricted above ovary, perianth throat and segments rose-pink, segments 11–17 mm long, rotate, outer segments 3–5 mm wide, inner 5–7 mm; staminodia erect, 5–8 mm long, middle portion of margin half

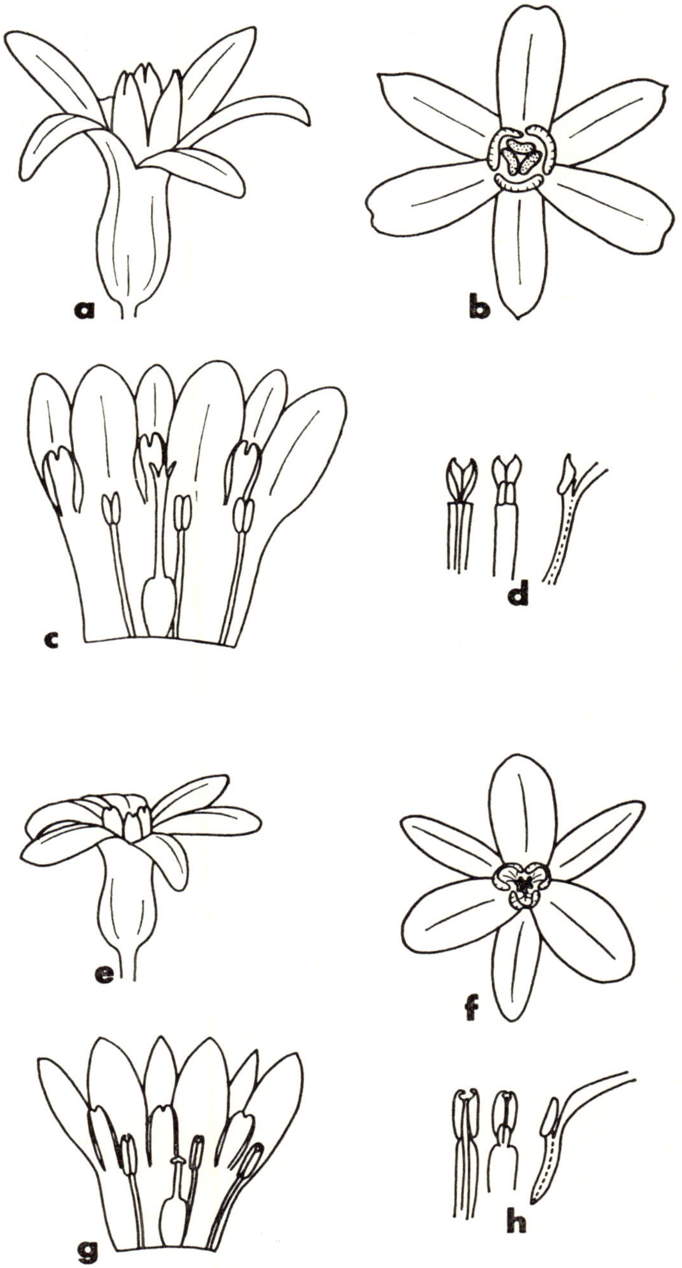

Fig. 16. Floral morphology of *Brodiaea pallida: a,* side view; *b,* top view; *c,* internal arrangement; *d,* stamens. *B. minor: e,* side view; *f,* top view; *g,* internal arrangement; *h,* stamen. (*a–c* and *e–g*, × 1½; *d* and *h*, × 2)

involuted, tip bilobed, white, approximate to stamens; stamens 3–5 mm below apex of staminodia, filament 0.5–4.0 mm long, with narrowly linear wing on abaxial surface, anther 1–5 mm long, rounded notch at apex with apical tips of anther lobes clawed inward; stigma 2–3 mm below anther apex, style 5–6 mm long, ovary globose, 3–4 mm long; in fruit perianth tube

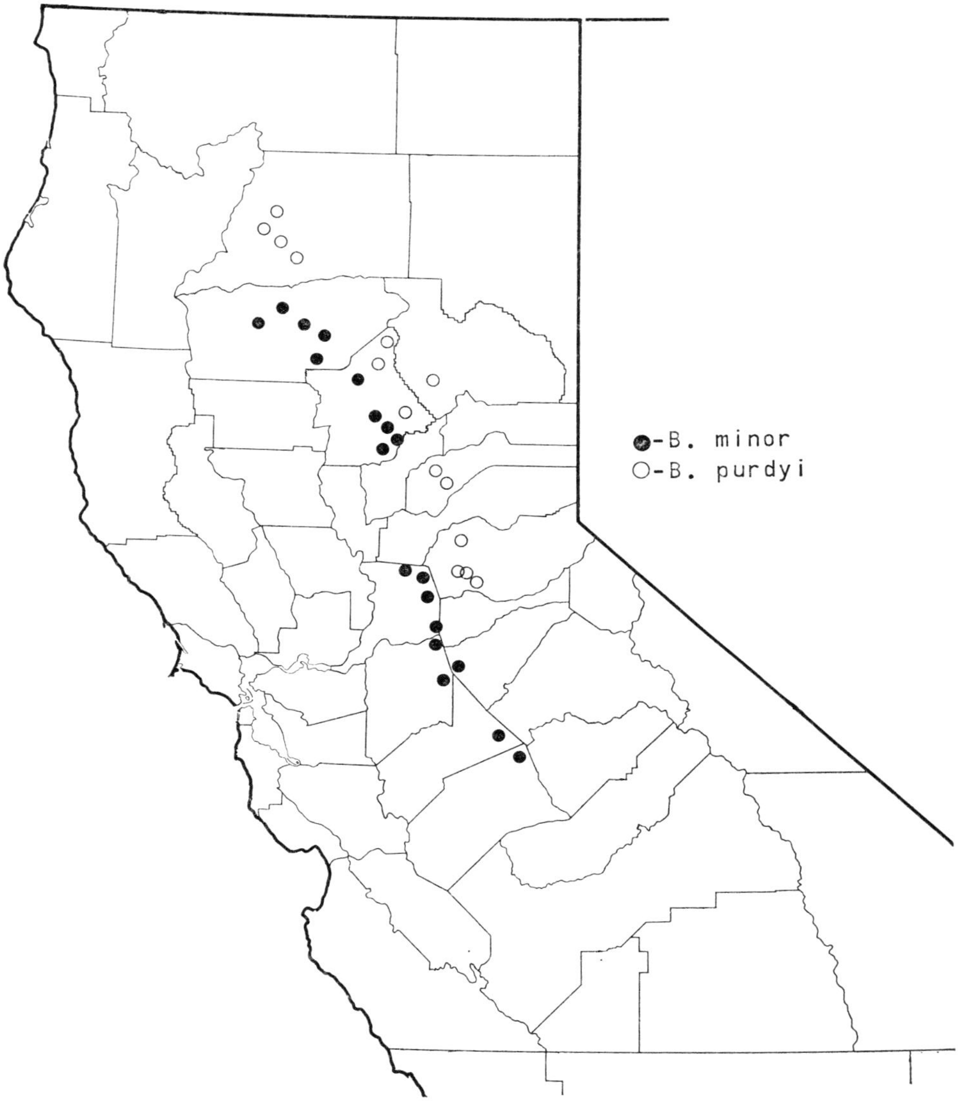

Fig. 17. Distribution of *Brodiaea minor*, *B. purdyi*.

opaque and highly resistant to splitting. Chromosome number n = 6 (Burbanck 1941), (fig. 16).

Distribution: (fig. 17) California, lowest foothills along E side of Central Valley from Tehama Co. to Merced Co. Flowers April to early May.

Community: Valley Grassland, in gravelly clay soils. Elevation 100–1,500 feet.

Representative specimens: CALIFORNIA. Tehama Co.: SE of Red Bluff, *Crum 1435;* 6 miles S of Corning, *Hoover 2298:* 0.8 mile E of Balls Ferry, *Niehaus 639;* 0.3 mile SE on Lane Valley road from Tehama Co. road A6, *652.* Butte Co.: 10 miles N of Chico, *Heller 14364;*

10.8 miles N of Oroville, *Balls & Lenz 20483;* 5 miles SW of Pentz, *Heller 11321;* 2.5 miles NE of Honcut, *Embree 45.* Sacramento Co.: Del Paso, *Hoover 1133;* Rio Linda, *2205;* 1.75 miles SE of Folsom, *Nordstrom 34;* Slough House, *Hoover 5297.* Amador Co.: 3 miles W of Ione, *2409.* San Joaquin Co.: 4 miles N of Clements, *Mason 4452;* Farmington, *Hoover 1056.* Calaveras Co.: 0.5 mile W of Burson, *Niehaus 517.* Stanislaus Co.: Warnerville, *Hoover 1048.* Merced Co.: 5 miles N of Snelling, *2065.*

12. **Brodiaea purdyi** Eastw., Proc. Calif. Acad. II. 6:427, pl. 58. 1896

From Colfax to hills back of Chico, 1896, *Purdy.* Holotype, CAS. *Hookera purdyi* Heller, Muhl. 6:83. 1910.

Corm 2–10 cm below ground level; leaves crescent-shaped in transection; scape 1.0–2.5 dm tall, pedicels 1–3 cm long; perianth tube 6–8 mm long, strongly constricted at throat, perianth throat and segments violet, segments 9–19 mm long, semi-rotate, outer segments 3–4 mm wide, inner 4–5 mm; staminodia erect, 6–8 mm long, margin three quarters involuted, tip bilobed, white, approximate to stamens; stamens 2–3 mm below apex of staminodia, filament 1–2 mm long, with narrowly triangular wing on abaxial surface, V-shaped notch between tips of anther lobes; stigma level with anther apices, style 7–9 mm long, ovary ovoid, 4–5 mm long; in fruit perianth tube opaque and highly resistant to splitting. Chromosome number n = 12, 24 (fig. 18).

Distribution: (fig. 17) California, upper foothills along E side of Central Valley from Shasta Co. to El Dorado Co. Flowers in June.

Community: Foothill Woodland, in open areas, usually on serpentine. Elevation 400–2,000 feet.

Representative specimens.—CALIFORNIA. Shasta Co.: 3 miles N of Redding, *Hoover 1191;* 5 miles E of Redding, *1209;* 0.5 mile N of Sheridan Creek, *Johansen 105.* Butte Co.: Butte Meadows, *Heller 11619;* 3 miles SW of Paradise, *Constance 2825;* 3 miles N of Cana, *Hoover 1173.* Plumas Co.: Prattville, *Combs;* Quincy Ranger Station, *Weatherby 1642.* Nevada Co.: Banner Hill, *Hall 10175;* E side of Wolf Mountain, *Niehaus 240.* El Dorado Co.: between Coloma and Rose Springs, *Brandegee;* 3 miles E of Rescue, *Robbins 1925;* 2 miles N of Shingle Springs, *Rose 52001;* White Rock Canyon, 3 miles NE of Placerville, *Belshaw 2336.*

Section Filifoliae Hoover, Am. Midl. Nat. 22:571. 1939

Flowers very small; perianth tube cylindric, in fruit becoming thin membranous and splitting, except in *B. kinkiensis;* perianth segments rotately spreading; staminodia reduced or lacking in one species.

13. **Brodiaea filifolia** S. Wats., Proc. Am. Acad. 17:381. 1882

San Bernardino, 1880, *Parish Brothers 362, B.* Holotype, GH.
Hookera filifolia Greene, Bull. Calif. Acad. 2:138. 1886.

Corm 5–10 cm below ground level; leaves subcrescent-shaped in transection; scape 2–3 dm tall, pedicels 1–4 cm long; perianth tube 6–8 mm long, perianth throat and segments rose-violet, segments 10–14 mm long, rotate, outer segments 3–5 mm wide, inner 5–7 mm; staminodia narrowly subulate, 2–4 mm long, reflexed over perianth segments; stamens exceeding staminodia 2–3 mm, filament 0.5–1.0 mm long, with triangular wing on abaxial surface; anther 3–5 mm long, broad rounded notch at apex; stigma 1–3 mm below anther apex, style 6–7 mm long, ovary globose, 4–5 mm long; in fruit perianth tube transparent and splitting. Chromosome number n = 12 (fig. 18).

Distribution: (fig. 19) California, local, San Bernardino Valley, Perris, Vista. Flowers April–May.

Community: Grassland, in clay soils of vernal pools. Elevation 300–1,000 feet.

Representative specimens: CALIFORNIA. San Bernardino Co.: Arrowhead Hot Springs, *Parish.* Riverside Co.: 1 mile E of Perris, *Roos 5667;* Vail ranch W of Murrieta, *Lenz 12226.* San Diego Co.: 4.5 miles S of Vista on Highway 395, *Gander 2053.*

Brodiaea filifolia is uncommon. It grows on clay soils at the edge of vernal

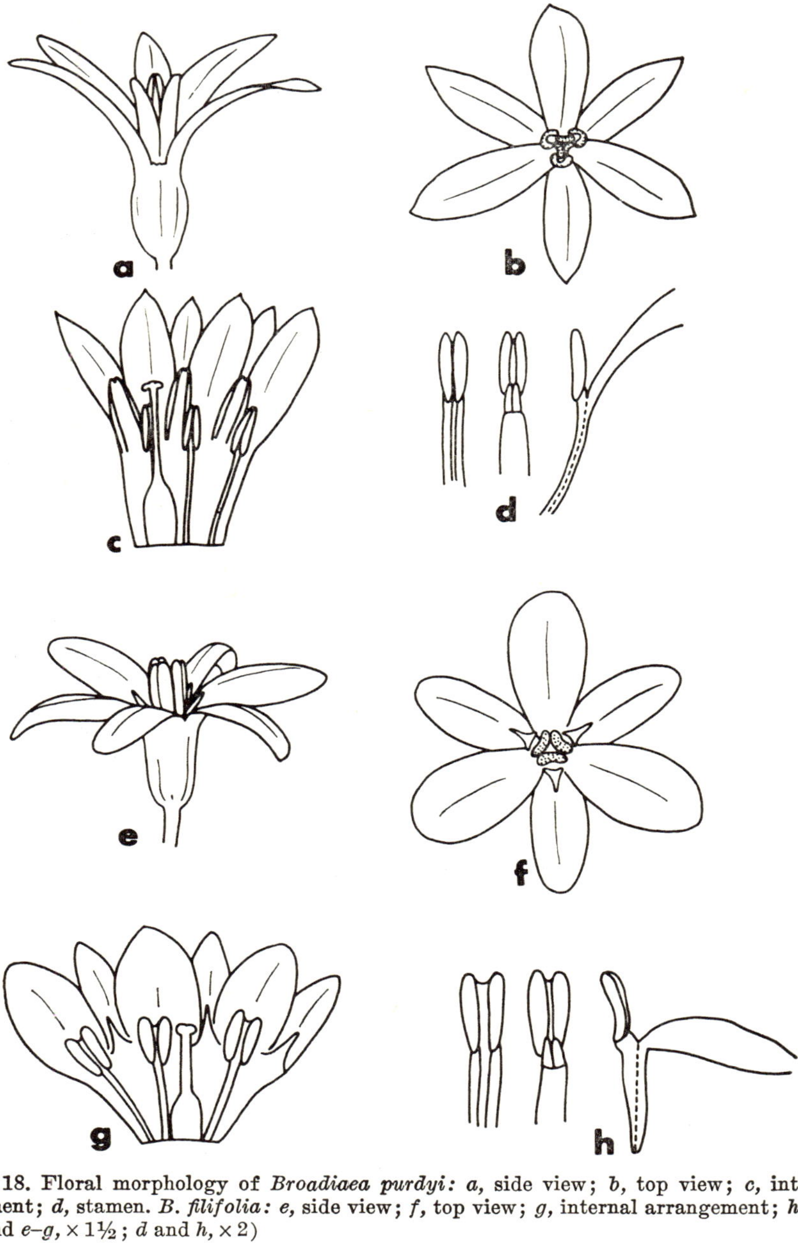

Fig. 18. Floral morphology of *Broadiaea purdyi: a,* side view; *b,* top view; *c,* internal arrangement; *d,* stamen. *B. filifolia: e,* side view; *f,* top view; *g,* internal arrangement; *h,* stamen. (*a–c* and *e–g*, × 1½; *d* and *h*, × 2)

pools or flood plains. The most extensive population, along the San Jacinto River bottom near Perris, Riverside County, is being encroached upon by cultivation.

14. **Brodiaea kinkiensis** Niehaus, Madroño 18:233. 1966

0.5 mile W of Stone at junction of road to Stone with main island axis road, San Clemente Island, Los Angeles Co., 1,395 feet, April 3, 1965, *Niehaus 407*. Holotype, UC.

Corm with heavy fibrous coat, 5–13 cm below ground level; single-leaved, terete with V-shaped groove in transection; scape 2–3 cm tall, pedicels 3–8 cm long; perianth tube 9–12 mm long, perianth throat and segments violet, segments 12–18 mm long, rotate, outer segments 5–7 mm wide, inner 8–11 mm; staminodia erect and well apart from stamens, 3–7 mm long, apically cuspidate; stamens slightly below apical level of staminodia, filament 1.5–2.0 mm long, with triangular wing on abaxial surface, anther 4–7 mm long, broadly notched at apex; stigma 1–2 mm below apex of staminodia, style 5–7 mm long, ovary ovoid, 6–9 mm long; in fruit perianth tube opaque and not splitting. Chromosome number n = 16 (Niehaus, 1965 as *B. filifolia*), (fig. 20).

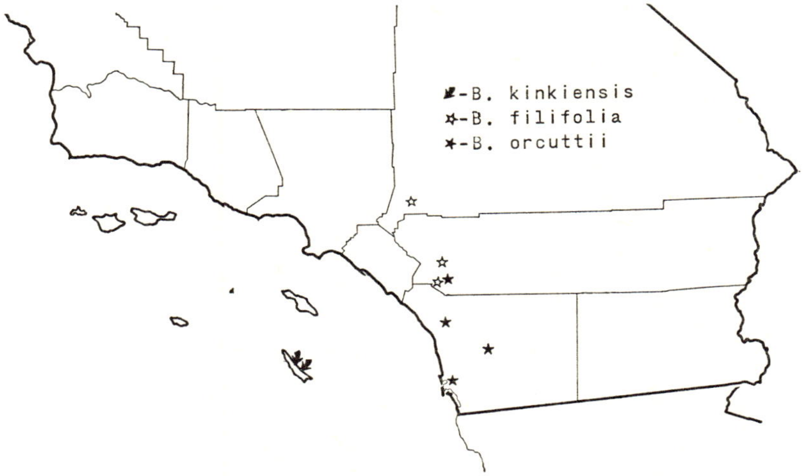

Fig. 19. Distribution of *Brodiaea kinkiensis, B. filifolia,* and *B. orcuttii.*

Distribution: (fig. 19) California, mesa top of San Clemente Island. Flowers May to June.
Community: Grassland, in clay flats along mesa.
Representative specimens: CALIFORNIA. Los Angeles Co.: 0.5 mile W of Stone, San Clemente Island, *Niehaus 407;* near Lemmon Tank, *397;* head of canyon N of Boulder, *Raven 17734;* mesa summit, *Murbarger 219.*

15. **Brodiaea orcuttii** (Greene) Baker, Gard. Chron. III. 20:214. 1896

Hookera orcuttii Greene, Bull. Calif. Acad. 2:138. 1886. Near San Diego and 30 miles to north, 1884, *C. R. Orcutt.* Holotype missing, apparently lost in San Francisco earthquake and fire of 1906. Lectotype, GH.

Hookera multipedunculata Abrams, Bull. Torrey Club 32:537. 1905. Cuyamaca Lake, *Abrams 3897.* Holotype, US. Isotypes, CAS, DS, GH, UC.

Brodiaea filifolia var. *orcuttii* Jepson, Fl. Calif. 1:288. 1922.

Corm 5–10 cm below ground level; leaves nearly terete with wide channel in transection; scape 0.8–2.5 dm tall, pedicels 1–5 cm long; perianth tube 3–5 mm long, perianth throat and segments rose-violet, segments 12–19 mm long, rotate, outer segments 3–5 mm wide, inner 5–7 mm; staminodia absent; stamen filament 4–6 mm long, anther 4–6 mm long, V-shaped notch at apex; stigma level with apex of anthers, style 7–11 mm long, ovary subovoid, 4–6 mm long; in fruit perianth tube transparent and splitting. Chromosome number n = 12 (Lenz, 1966), (fig. 20).

Distribution: (fig. 19) California, southern Riverside Co. to San Diego Co. Flowers in May.
Community: Valley Grassland, near intermittent streams and vernal pools. From sea level to 5,000 feet.
Representative specimens: CALIFORNIA. Riverside Co.: vernal pools W of Murrieta, Lathrop. San Diego Co.: San Marcos, *Gander;* Henshaw Dam, *M. E. Jones;* Cuyamaca Lake, *Abrams*

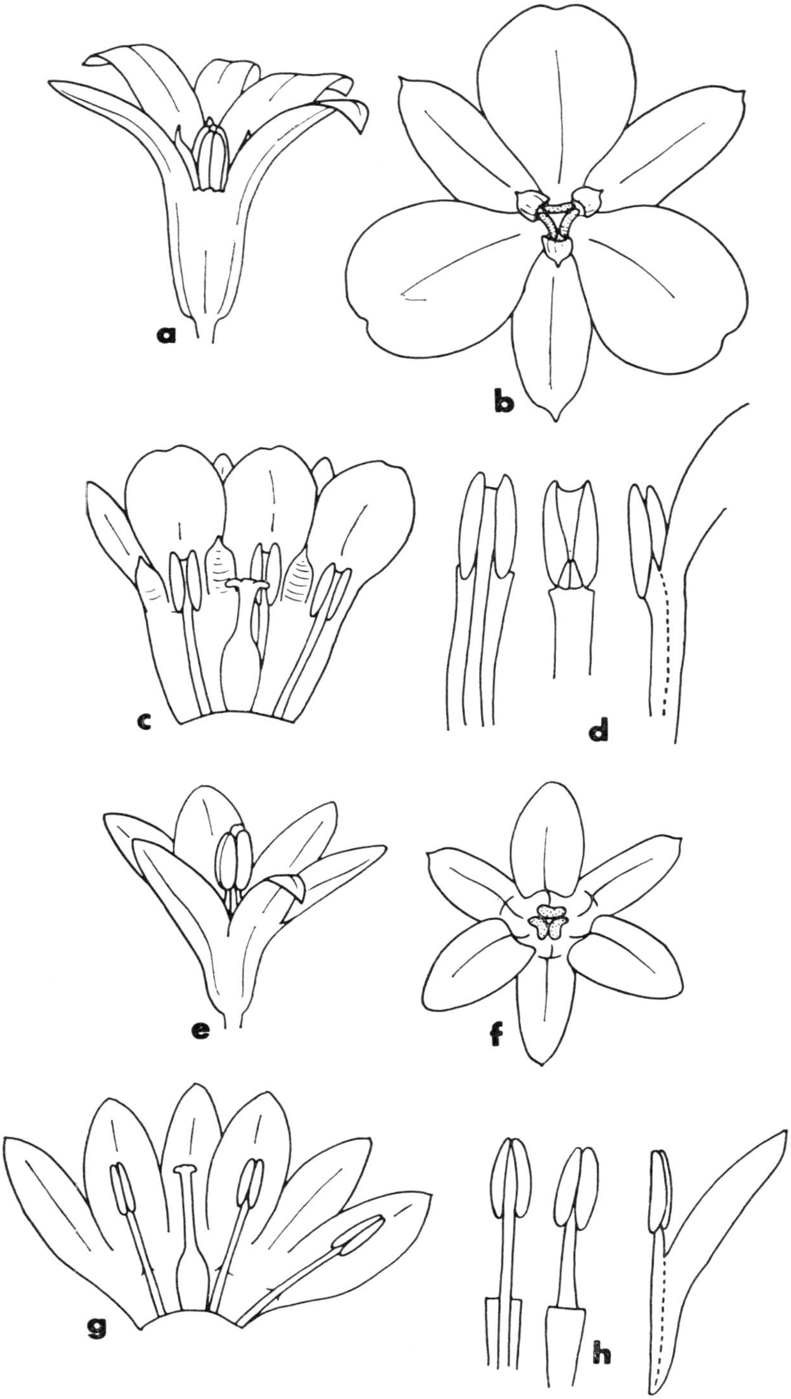

Fig. 20. Floral morphology of *Brodiaea kinkiensis*: *a*, side view; *b*, top view; *c*, internal arrangement; *d*, stamen. *B. orcuttii*: *e*, side view; *f*, top view; *g*, internal arrangement; *h*, stamen. (*a–c* and *e–g*, × 1½; *d* and *h*, × 2)

3897; Linda Vista on Kearney Mesa, *Gander 8315;* Woodwardia Canyon, Otay Mountain, *Wolf 7971.*

Brodiaea orcuttii is found along the margins of intermittent streams and vernal pools. At one time it formed an extensive population on Kearney Mesa north of San Diego, but this area is now covered with tract housing, and *B. orcuttii* occurs only on undisturbed portions of military reservations on the mesa.

APPENDIX

List of voucher specimens used in this study. Collection numbers are the authors' and in California unless stated otherwise. Deposited at University of California, Berkeley.

Brodiaea appendiculata Hoover
667. Eagle Nest road, Sacramento Co. n = 6.
515. 1 mile W of Ione, Amador Co. n = 6.
516. 5 miles S of Ione, Amador Co. n = 6.
635. 5 miles E of Plainada, Merced Co. n = 6.

Brodiaea californica Lindl.
651. 1.5 miles W of Bear Creek Bridge, Hwy 44, Shasta Co. n = 6.
244. Wolf Mountain, Nevada Co. n = 6.
245. Pet Hill, Nevada Co. n = 6.
543. 1.5 miles SE of Nevada-Yuba Co. line, Hwy 20. Nevada Co. n = 6.

Brodiaea leptandra (Greene) Baker
679. 2.6 miles W of Napa-Sonoma Co. line on Spring Mountain road, Sonoma Co. n = 12.
680. 3.9 miles W of Pope Valley on Ink Grade road, Napa Co. n = 12.
UCBG.* 55.926. N of Mt. St. Helena, Lake Co. 2n = 12.

Brodiaea coronaria (Salisb.) Engler subsp. *coronaria*
706. Jenny Creek, Oregon. n = 6.
708. Keno, Oregon. n = 12.
512. 4 miles W of Ione, Sacramento Co. n = 12.
634. Slough House road, Sacramento Co. n = 12.
741. Oroville, Butte Co. n = 18.
Keator MD 9. N of Canby, Modoc Co. n = 6.
Taylor, G. Victoria, British Columbia. n = 12.

Brodiaea coronaria subsp. *rosea* (Greene) Niehaus
556. Indian Valley, Lake Co. n = 6.

Brodiaea elegans Hoover
128. Grass Valley, Nevada Co. n = 8.
709. Nevada City, Nevada Co. n = 8.
552. Auburn Airport, Placer Co. n = 8.
668. 2 miles SE of Folsom, El Dorado Co. n = 8.
670. 1.5 miles N of U.S. 50 on Lotus road, El Dorado Co. n = 8.
236. 3 miles E of Sonora, Tuolumne Co. ("var. *mundula*"). n = 6, 8, 12, 16.
828. Mist Falls, S fork Kings River, Fresno Co. n = 8.
695. Grants Pass, Oregon. n = 16.
571. Gasquet, Del Norte Co. n = 16.
566. Alderpoint, Humboldt Co. n = 16.
567. 2 miles E of Bridgeville, Humboldt Co. n = 16.
561. Ukiah, Mendocino Co. n = 16.
525. Bartlett Springs, Lake Co. n = 16.
692. Santa Rosa, Sonoma Co. n = 16.
522. Yountville, Napa Co. n = 16.
584. S of San Jose, Santa Clara Co. n = 16.
546. Indian Springs, Nevada Co. n = 16.
513. 3.5 miles west of Highway 104 from Sacramento/Amador Co. line, Sacramento Co. n = 16.
237. Near Vallecito, Calaveras Co. n = 16.
232. Chinese Camp, Tuolumne Co. n = 16.
685. Soulsbyville, Tuolumne Co. n = 16.
761. 2 miles E of Sonora, Tuolumne Co. n = 16.
767. Woods Creek S of Chinese Camp, Tuolumne Co. n = 16.

* Univ. Calif. Botanical Garden.

677. Three Rivers, Tulare Co. n = 16.

Brodiaea elegans subsp. *hooveri* Niehaus

696. McKenzie River near Eugene, Oregon. n = 20.
702. Table Rock road near Medford, Orgeon. n = 20.

Johnston, L. D. 8 miles S of Corvallis, Oregon. n = 20.

Brodiaea terrestris Kellogg subsp. *terrestris*

364. Crystal Lake, San Mateo Co. n = 6.
674. Havens Neck, Mendocino Co. 2n = 36.
535. Drakes Beach, Marin Co. 2n = 36.
510. Jolon, Monterey Co. n = 18.

Roderick, W. San Simeon, San Luis Obispo Co. 2n = 36.

Brodiaea terrestris subsp. *kernensis* (Hoover) Niehaus

659. Terra Bella, Tulare Co. n = 24.
746. E of Fountain Springs, Tulare Co. 2n = 48.
747. E of Ducor, Tulare Co. 2n = 48.
632. Garces Monument, Kern Co. 2n = 48.
743. Bakersfield dump, Kern Co. 2n = 48.

Brodiaea jolonensis Eastw.

511. Jolon, Monterey Co. n = 6, 2n = 12.
509. Isla Vista, Santa Barbara Co. 2n = 36.
508. E of Santa Barbara, Santa Barbara Co.
662. Otay Mesa, San Diego Co. 2n = 36.

Brodiaea stellaris S. Wats.

673. Harrison Grade, Sonoma Co. n = 6.

Keator, UCBG. Bridgeville, Humboldt Co. n = 6.

Brodiaea insignis (Jepson) Niehaus

753. 0.1 mile N of Salt Creek fire road on South Fork of Kaweah road, Tulare Co.
754. 5.5 miles S from old town of Three Rivers, Tulare Co. 2n = 32.
755. 4.7 miles S from old town of Three Rivers, Tulare Co.

Brodiaea pallida Hoover

526. W of Chinese Camp, Tuolumne Co. n = 6, 2n = 12.

Brodiaea minor (Benth.) S. Wats.

639. 0.8 mile E of Balls Ferry, Tehama Co. n = 6.
653. Lane Valley road, Tehama Co.
655. Pentz road E of Oroville, Butte Co. 2n = 12.
517. 0.5 mile W of Burson, Calaveras Co. n = 6.

Brodiaea purdyi Eastw.

648. 3 miles E of Redding, Shasta Co. 2n = 48.
658. Table Mountain, Butte Co. 2n = 48.
240. E side of Wolf Mountain, Nevada Co. 2n = 24, n = 12.
528. 1 mile N of Shingle Springs, El Dorado Co. 2n = 24, n = 12.

Brodiaea filifolia S. Wats.

661. 1 mile E of Perris, Riverside Co. n = 12.

Brodiaea kinkiensis Niehaus

407. 0.5 mile W of Stone, San Clemente Island, Los Angeles Co. n = 16, 2n = 32.
397. Near Lemmon Tank, San Clemente Island, Los Angeles Co.

Brodiaea orcuttii (Greene) Baker

663. N of Miramar Naval Air Station, San Diego Co. n = 12.

LITERATURE CITED

ALSTON, R. E., and B. L. TURNER

1963. Biochemical systematics. Prentice-Hall, Englewood Cliffs, N.J.

BABCOCK, E. B.

1947. The genus *Crepis* I. The taxonomy, phylogeny, distribution, and evolution of *Crepis*. Univ. Calif. Publ. Bot. 21:1–198.

BAKER, J. G.

1871. Revision of the genera and species of herbaceous capsular gamophyllous Liliaceae. Jour. Linn. Soc. 11:349–436.

1896. The genus *Brodiaea* and its allies. Gard. Chron. ser. 3, 20:213–214.

BELLING, J.

1926. The iron-acetocarmine method of fixing and staining chromosomes. Biol. Bull., 50: 160–162.

BENTHAM, G.

1857. Plantae Hartwegianae.

BRITTEN, J.

1886. Hookera vs. Brodiaea. Jour. Bot. 24:49–51.

BURBANCK, M. P.

1941. Cytological and taxonomic studies in the genus *Brodiaea*. Bot. Gaz. 103:247–265.

CRANWELL, L. M.

1953. New Zealand pollen studies: Monocots. Harvard Univ. Press.

DAVIS, P. H., and V. H. HEYWOOD

1963. Principles of angiosperm taxonomy. D. Van Nostrand Co., Inc., New York. Inc., New York.

EASTWOOD, A.

1896. Description of some new species of California plants. Proc. Calif. Acad. ser. 2, 6:422–430.

1938. New species in Liliaceae. Leafl. West. Bot. 2:109–112.

ELLISON, W. L., R. E. ALSTON, and B. L. TURNER

1962. Methods of presentation of crude biochemical data for systematics purposes with particular reference to the genus *Bahia* (Compositae). Am. Jour. Bot. 49:599–604.

ENGLER, A.

1887. Lilaceae *in* Engler and Prantl, Die naturlichen Pflanzenfamilien. part 2, 5:10–91.

1899. Handliste, zu kultivierenden Liliifloren. Notizbl. Konigl. Bot. Gart. 2:309–317.

ERDTMAN, G.

1952. Pollen morphology and plant taxonomy: Angiosperms. Almquist and Wiksell, Stockholm.

ESAU, K.

1953. Plant anatomy. John Wiley and Sons, New York.

FAVARGER, C. L., and J. CONTANDRIOPOULOUS

1961. Essai sur l'endemisme. Bul. Soc. Bot. Suisse 71:384–408.

FOSTER, A. S.

1949. Practical plant anatomy. D. Van Nostrand Co., New York.

GEIGER, M. J. T.

1929. A study of the origin and development of the Cormlet of Gladiolus. Catholic Univ. of America, Washington, D.C.

GEISSMAN, T. A.

1962. The chemistry of the flavonoid compounds. Oxford Press, London.

GRANT, V.

1959. Natural history of the Phlox family. Systematic botany. Vol. 1. Martinus Nijhoff, the Hague.

GREENE, E. L.

1886. Studies in the botany of California and parts adjacent. I. Some genera which have been confused under the name *Brodiaea*. Bull. Calif. Acad. 2:125–144.

1887. Miscellaneous species, new or rare. Pittonia 1:60–74.

HARBORNE, J. B.

1964. Biochemistry of phenolic compounds. Academic Press, London and New York.

HARVEY, P. M., and W. F. GRANT

1965. A polygonal presentation of chromatographic investigations on the phenolic content of certain species of Lotus. Canad. Jour. Genet. and Cytol. 7:40–51.

HELLER, A. A.

1903. Notes on plants from middle western California. Bull. So. Calif. Acad. 2:65–66.

1910. New combinations. III. Muhl. 6:83.

HOOVER, R. F.

1936. New California plants. Leafl. West. Bot. 1:225–230.

1937. A new California species of *Brodiaea.* Madroño 4:130–132.

1938. New California plants. Leafl. West. Bot. 2:128–133.

1939*a*. A definition of the genus *Brodiaea.* Bull. Torr. Bot. Club. 66:161–166.

1939*b*. A revision of the genus *Brodiaea.* Am. Midl. Nat. 22:551–574.

1955. Further observations on *Brodiaea* and some related genera. Plant Life 11:13–22.

1957. Observation on California plants. IV. Leafl. West. Bot. 8:129–133.

HUTCHINSON, J.

1959. The families of flowering plants. II. Monocotyledons. Second edition. Clarendon Press, Oxford.

JAHNS, R. H.

1954. Investigations and problems of southern California geology. 5–29. In Geology of Southern California. Bull. 170. Division of Mines, San Francisco.

JEPSON, W. L.

1922. Flora of California, Vol. 1. Univ. of Calif. Press, Berkeley.

1925. A manual of the flowering plants of California. Univ. of Calif. Press, Berkeley.

JOHANSEN, D. A.

1932. The chromosomes of the California Liliaceae. I. Am. Jour. Bot. 19:779–783.

KEATOR, G.

1967. Ecological and taxonomic studies on the genus *Dichelostemma.* Ph.D. thesis, Univ. of Calif. Library, Berkeley.

KELLOGG, A.

1859. *Brodiaea terestria.* Proc. Calif. Acad. 2:6–7.

KIMBER, G., and R. RILEY

1963. Haploid angiosperms. Bot. Rev. 29:480–531.

KRUCKEBERG, A. R.

1954. The ecology of serpentine soils. III. Plant species in relation to serpentine soils. Ecology 35:267–274.

1957. Variation in fertility of hybrids between isolated populations of the serpentine species, *Streptanthus glandulosus* Hook. Evolution 11:185–211.

KUNTH, C. S.

1843. Enumeratio Plantarum. Vol. 4:469.

LENZ, L. W.

1966. Chromosome numbers in the Allieae (Liliaceae). Aliso 6:81–82.

LEWIS, H.

1953. Chromosome phylogeny and habitat preference in *Clarkia.* Evolution 7:102–109.

LEWIS, H., and C. EPLING

1959. *Delphinum gypsophilum,* a diploid species of hybrid origin. Evolution 13:511–525.

LEWIS, H., and M. E. LEWIS

1955. The genus *Clarkia.* Univ. Calif. Publ. Bot. 20:241–392.

MASON, H. L.

The 1946*a*. The edaphic factor in narrow endemism. I. The nature of environmental influences. Madroño 8:209–226.

1946*b*. Ibid. II. The geographic occurrence of plants of highly restricted patterns of distribution. Madroño 8:241–257.

MOORE, H. E.
1953. The genus Milla and its allies. Gent. Herb. 8:262–294.
MUNZ, P. A.
1959. A California flora. Univ. of Calif. Press, Berkeley.
NIEHAUS, THEODORE F.
1965. *in* Documented chromosome numbers of plants. Madroño 18:123.
1966. A new species of *Brodiaea* from San Clemente Island, California. Madroño 18:233–235.
ORNDUFF, R.
1969. The systematics of populations in plants. *In* Systematic Biology, Natl. Acad. Sci. Publ. 1692. Washington, D.C.
RAVEN, P. H., and H. J. THOMPSON
1964. Haploidy and angiosperm evolution. Am. Nat. 98:251–252.
SALISBURY, R. A.
1808. Paradisus Londonensis. Pl. 98.
SCORA, R. W.
1967. Interspecific relationships in the genus monarda (Labiatae). Univ. Calif. Publ. Bot. 41.
SECOND INTERN. BOT. CONGRESS
1906. Nomina Conservanda 1053. Verhandlungen des Internationalen Botanischen Kongresses in Wien. Verlag von Gustav Fischer, Jena.
SMITH, F. H.
1930. The corm and contractile roots of *Brodiaea lactea.* Am. Jour. Bot. 17:916–927.
SMITH, J. E.
1811. Characters of a new Liliaceous genus called *Brodiaea.* Trans. Linn. Soc. London 10:1–5.
STEBBINS, JR., G. L.
1950. Variation and evolution in plants. Columbia Univ. Press, New York.
1966. Chromosomal variation and evolution. Science 152:1463–1469.
STEBBINS, G. L., and J. MAJOR
1965. Endemism and speciation in the California flora. Ecol. Monogr. 35:1–35.
SWAIN, T.
1966. Comparative phytochemistry. Academic Press, London and New York.
SWANSON, C. P.
1960. Cytology and cytogenetics. Prentice-Hall, Englewood Cliffs, New Jersey.
THOMPSON, H. J.
1960. A genetic approach to the taxonomy of *Mentzelia lindleyi* and *M. crocea* (Loasaceae). Brittonia 12:81–93.
TING, W. S.
1966. Determination of Pinus species by pollen statistics. Univ. Calif. Publ. Geol. Sci. 58.
TORREY, J.
1857. Pacific Railroad Report 4:149.
TRAUB, H. P.
1954. The genus Nothoscordum: diagnosis, key to species and synonymy. Plant Life 10:123–127.
TRAUB, H. P., and H. MOLDENKE
1955. The genus Ipheion: diagnosis, key to species and synonymy. Plant Life 11:125–130.
TURNER, H. W.
1894. Description of the gold belt (Sierra Nevada, Calif.). *In* U. S. Geol. Survey, Atlas of the United States, Jackson Folio 11.
WATSON, S.
1879. Revision of the North American Liliaceae. Proc. Am. Acad. 14:213–288.
WOOD, A.
1868. A sketch of the natural order Liliaceae. Proc. Acad. Phila. 20:165–174.
YAUSY, L.
1942. The floral morphology of several species of *Brodiaea.* M.A. thesis. Stanford Univ. Library.

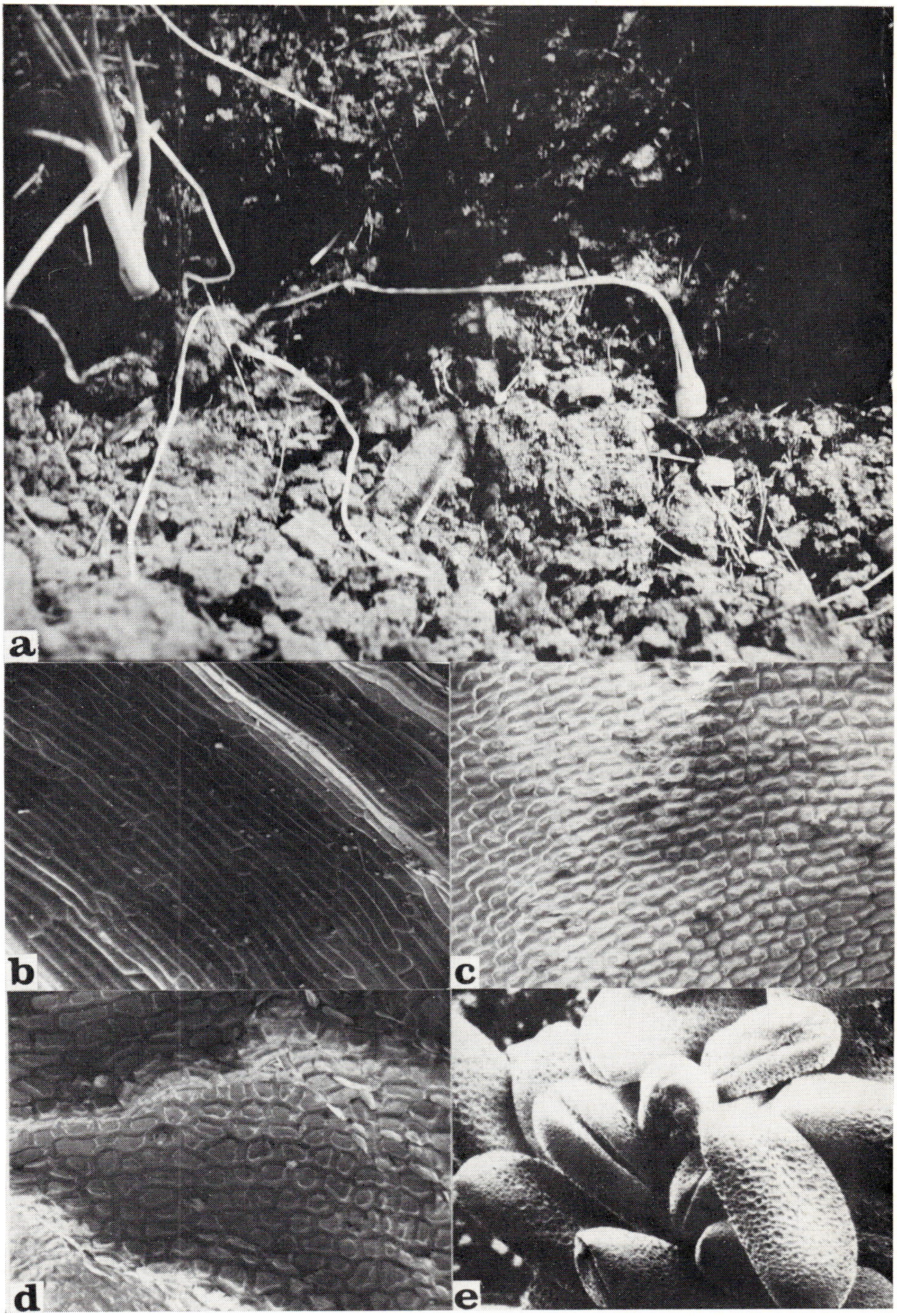

Plate 1. *a*. Plant of *B. leptandra* with soil removed to show parent corm and cormlet connected by what appears to be a contractile root. *b*. Seed coat of *Brodiaea*. *c*. Seed coat of *Dichelostemma*. *d*. Seed coat of *Triteleia*. *e*. Pollen of *Brodiaea*. Scanning electron micrographs *b, c, d,* 90 X; *e,* 300 X. Prepared in Electronics Research Laboratory, University of California, Berkeley, under grants NSF GB-6428 and NIH GM-15536.